Intelligent Micro Motion Device (Micromouse) Technology and Application Series
Research Result of Tianjin "the Belt and Road" Joint Laboratory (Research Center)
Engineering Practice Innovation Project (EPIP) Teaching Mode Planned Textbook

Micromouse Design Principles and Production Process
(Elementary)

Compiled by Wang Chao, Gao Yi, Song Lihong
Translated by Wang Xiaoqin, Yan Jingyi, Zhou Fanyu

U0183849

中国铁道出版社有限公司
CHINA RAILWAY PUBLISHING HOUSE CO., LTD.

Introduction to the contents

The book is bilingual in both Chinese and English, based on the TQD-Micromouse-JQ provided by Tianjin Qicheng Science and Technology Co., Ltd., which is the elementary-level of a series of books on Micromouse Technology and Application.

The book is based on real engineering projects, through "Elementary Knowledge", "Comprehensive Practice", "Advanced Skills and Competitions" and "Extended Application". All of these four chapters are describe the development, hardware, development environment, and function debugging of the Micromouse; interactive control and attitude control of the Micromouse; analysis of the Micromouse algorithm; advanced functions; application expansion of micromouse technology, etc. The appendix of this book provides the relevant knowledge of the international Micromouse maze competition, such as the device list of TQD-Micromouse-JQ, the Micromouse maze library, bilingual comparison table of glossary, and the international curriculum standard.

The book is equipped with a wealth of resources such as videos, pictures, texts, etc. on important knowledge points, skills points and literacy points. The reader can obtain relevant information by scanning the QR code in the book.

The book is suitable as the textbook of basic education school to carry out vocational enlightenment, scientific and technological activities and characteristic education. It is also suitable as the textbook of vocational education. It can also be used as a training book for relevant engineering and technical personnel or reference book for Micromouse lovers.

Wang Chao

Wang Chao is currently a Professor in School of Electrical and Information Engineering, Tianjin University, China. He is a member of the Teaching Guiding Committee for Automation Majors under the Ministry of Education of China. His current research interests include multiphase flow measurement and instrumentation, electrical tomography (ERT, ECT, EMT and EST). His courses at Tianjin University include computer control technology and industrial control networks. Since 2010, he has introduced Micromouse as an important carrier of practical teaching into school of electrical and information engineering for the first time. Two teams of Tianjin University won the first and second place at 2018 APEC Micromouse competition.

Gao Yi

Gao Yi is a master supervisor of School of Electronic Information Engineering at Nankai University, deputy director of Electronic Information Experimental Teaching Center, deputy director of Youth Working Committee of Tianjin MCU Society and member of the judging group of many technology competition of college students and Tianjin vocational skills competition. He has participated in "national high-tech research and development plan (863 plan) projects", "Tianjin science supporting plan key projects" and multiple horizontal scientific research projects. He has led teams to participate in the national undergraduate electronic design competition, Tianjin electronic design competition, Tianjin internet of things competition, Tianjin IEEE Micromouse competition for college students, APEC international Micromouse competition and national robot competition.

Song Lihong

Song Lihong is general manager of Tianjin Qicheng Science and Technology Co., Ltd. and founder of Qicheng Micromouse. The company has been committed to the research and development, design, production, promotion and service work of teaching instruments about embedded system, internet of things and artificial intelligence used in higher education, vocational education and basic education. Qicheng has sponsored the "Qicheng Cup" Micomouse maze competition of college students and the intelligent micro motion device (Micromouse) competition of vocational college skills competition more than 40 times. Since 2016, the company has been actively engaged in the technical support service of international project Luban Workshop. As an innovative educational equipment in China, Qicheng Micromouse has been brought to Thailand, India, Indonesia, Pakistan, Cambodia, Nigeria, Egypt and other countries and has been favored by teachers and students in these countries. Qicheng Micromouse has made contributions to "the Belt and Road" initiative.

Wang Xiaoqin

Wang Xiaoqin is the first class teacher of Tianjin Dongli District Vocational Education Center, deputy director of processing and manufacturing department and district-level disciplinary leading figure. She has won the first prize of Tianjin Vocational College Teachers' skills competition in 2008, the second prize of Tianjin education and teaching achievements in 2014, the first prize of Tianjin informatization teaching competition in 2016, the second prize of the national lecture contest in 2016, and the first prize of the second national micro class competition in 2016. She has led the team to participate in many contests both home and abroad, including the electronic circuit debugging skills competition, IEEE Micromouse competition and APEC international Micromouse competition.

Yan Jingyi

Yan Jingyi, assistant to the general manager of Tianjin Qicheng Science and Technology Co.,Ltd., studied in the University of California, Santa Cruz. In 2015, she served as the accompaniment interpreting for professor MIT David Otten, chairman of the organizing committee of APEC Micromouse international competition. In 2018, she served as the English translator and simultaneous interpreter for Mr. Davaanyam, vice chairman of the board of directors of new Mongolia Education Group when he visited in Tianjin. In 2018, she went to Cambodia as a volunteer interpreter for Bun Phearin, president of National Polytechnic Institute of Cambodia. She was helping Cambodia students learn the Chinese Micromouse technology at the same time. Yan makes contributions to the education in countries along "the Belt and Road" for a long time.

Zhou Fanyu

Zhou Fanyu, master of English education, University of Central Oklahoma, is currently working in the international exchange and cooperation office of Tianjin Vocational College of Mechanics and Electricity. During her stay in the United States, she participated in research projects such as "the Comparative Study of Chinese and English Languages". She taught a large number of non-American students whose English was the second language in Confucius Classroom, and gained a lot of teaching experience. In 2018, she participated in the summer Davos forum and served as the interpreter and affairs officer for Sun Xiansheng, secretary general of the International Energy Forum. In 2018, she participated in the establishment and construction of Portugal Luban Workshop. In 2018, she published a bilingual paper on "Research on the Training of Practical Ability in Micromouse Project by the Teaching Model of Luban Workshop". In 2019, she participated in the establishment and construction of Madagascar Luban Workshop. During this period, she completed the translation of several meetings, accompanying translation, and translated more than 200,000 words of documents and materials.

Micromouse is a micro, intelligent motion device (or embedded microrobot) composed of embedded micro controllers, sensors and electromechanical moving parts. Micromouse can reach the predetermined destination fast by automatically memorizing all the routes in various mazes and selecting the optimal path out with the aid of suitable algorithms. The Micromouse competition involves wide-ranging scientific knowledge such as mechatronics, cybernetics, optics, programming and artificial intelligence.

For more than 40 years, the Institute of Electrical and Electronics Engineers (IEEE) has hosted an annual Micromouse competition. Since its inception, the international event has featured extensive and active participation, especially that of students from colleges and universities in the US and Europe. Some universities even offer an elective course on the principles and production process of Micromouse. In 2007, Shanghai and some other cities in the Yangtze River Delta started to stage small-scale, experimental Micromouse competitions. In 2009, Tianjin Qicheng Science and Technology Co., Ltd. introduced the competition to Tianjin and added local features to create an updated version based upon the Engineering Practice Innovation Project (EPIP) teaching model. These pioneering efforts helped boost future Micromouse events and played a key role in integrating them into classroom teaching. Years of exploration and progress have made Micromouse competitions to serve as educational platforms that encourage innovation and practice. By bringing together expertise and interest, the multi-dimensional and pioneering competitions have been essential for cultivating students' capabilities in practice and innovation, reforming curriculum and improving education.

In order to further promote and apply the achievements of Micromouse, we specially organized the book of *Micromouse Design Principles and Production Process(Elementary)* for basic education and vocational education. This book is based on the TQD-Micromouse-JQ provided by Tianjin Qicheng Science and

Technology Co.,Ltd. As a teaching carrier, practice teaching is carried out from shallow to deep, and from easy to difficult, step-by-step teaching.

This book will teach users with basic principles of Micromouse and end up as a profession in the field by following a step-by-step method that is also used in compiling the book. After reading the book, users will acquire more knowledge about engineering practice, enrich their experience in technology application, open up a broader vision in expertise, and become more professional.This book's ultimate goal is to cultivate innovation-minded practitioners. The select cases in the book are all adapted from real-life engineering projects. Also all the authors are came from enterprises, colleges and universities that have long remained committed to R&D on Micromouse or that have won awards in international competitions.

This book offers abundant videos, pictures and texts, etc.on important knowledge points, skills points and literacy points. The reader can scan the QR code in the book to get these supporting resources. The authors' rich international teaching experience has made the book became a practical teaching carrier to promote international talent training. Institutions of higher education and vocational colleges can use the book to guide their practicums on innovation. Intelligent micro motion device (Micromouse) technology and application series are the research result of Tianjin "the Belt and Road" Joint Laboratory (Research Center)— Tianjin Sino-German and Cambodia Intelligent Motion Device and Communication Technology Promotion Center and are also the Engineering Practice Innovation Project (EPIP) teaching model planned textbook. The book is suitable as the textbook of basic education school to carry out vocational enlightenment, scientific and technological activities and characteristic education. It is also suitable as the textbook of vocational education. It can also be used as a training book for relevant engineering and technical personnel or reference book for Micromouse lovers.

The appendix of this book provides the international training course standard of "Micromouse Design Principles and Production Process" (applicable to secondary vocational colleges). There is a total of 60 hours of teaching content. The course design and implementation adopt a method of integrating theory and practice, combining brain and hands. Throughout the technical skills training, it serves for the internalization of professional ethics. The content of this course is highly integrated with the "Luban Workshop" construction projects in many countries. Serving the "the Belt and Road" initiative, spreading Chinese vocational education standards, providing rich practical teaching resources for countries along "the Belt and Road" route, serving the training of skilled personnel in various fields.

This book is co-authored by Wang Chao, professor of Tianjin University; Gao Yi, associate professor of Nankai University; and Song Lihong, general manager of Tianjin Qicheng Science and Technology Co.,Ltd., the founder of Qicheng Micromouse. The English version was translated and compiled by Wang Xiaoqin, first class teacher of Tianjin Dongli District Vocational Education Center, Yan Jingyi, a general manager assistant of Tianjin Qicheng Science and Technology Co., Ltd. and Zhou Fanyu, lecturer in Tianjin Vocational College of Mechanics and Electricity. Li Xin, senior teacher of Tianjin Dongli District Vocational Education Center, Liu Xuewen, senior teacher of Tianjin Dongli District Vocational Education Center, Dong Qinglin, senior teacher of Tianjin Dongli District Vocational Education Center, Li Jun, first class teacher of Tianjin Dongli District Vocational Education Center, Zhang hongshu, senior teacher of Tianjin Dongli District Vocational Education Center and Li Xiaochen, an experimentalist of Electronic Information and Optical Engineering College in Nankai University, participated in the collation and translation of this book. David Otten, professor of Massachusetts Institute of Technology (MIT) in USA, Peter Harrison, professor of Birmingham City University in UK and António Valente, professor of University of Trás-os-Montes and Alto Douro in Portugal are proofreaders of the English version and they specially wrote congratulatory letters for this books. The book has received great and generous support from scholars and experts who come from Tianjin University, Nankai University, Tianjin Dongli District Vocational Education Center, Tianjin First Light Industry School, Tianjin Economics and Trade School, Massachusetts Institute of

Technology (MIT), Birmingham City University, and University of Trás-os-Montes and Alto Douro. Chen Likao, Qiu Jianguo and Song Shan, are the employees of Tianjin Qicheng Science and Technology Co.,Ltd. provided practical engineering cases, QR codes, videos and PPT course resources for the book. We also owe great gratitude to Tianjin Municipal Education Commission, China Railway Publishing House Co., Ltd. and Tianjin Qicheng Science and Technology Co.,Ltd. for their invaluable guidance and support. This book is sponsored to compiled by Tianjin Dongli District Vocational Education Center, published by China Railway Publishing House Co., Ltd. and will be used in countries along "the Belt and Road" route through the Luban Workshop program.

There will be some gaps or even errors in the book due to a tight publishing schedule and insufficient consideration from the authors, so any constructive criticisms and suggestions are greatly welcomed.

Authors
August, 2020

Contents

Chapter 1　Elementary Knowledge.................................001

Project 1　Evolution of Micromouse... 003

Task 1　Origins of Micromouse 003

Task 2　Competition and Debugging Environment of Micromouse........ 010

Project 2　Micromouse Hardware Structure 013

Task 1　Components of Micromouse 013

Task 2　The Control Structure of Micromouse 015

Project 3　Development Environment of Micromouse.................... 018

Task 1　Arduino Development Environment 018

Task 2　Logical Thinking of Micromouse 021

Project 4　Basic Function Control of Micromouse 026

Task 1　Micromouse Sees the World 026

Task 2　Micromouse Moves.. 031

Chapter 2　Comprehensive Practice.........................035

Project 1　Interaction Control of Micromouse............................ 037

Task 1　Human-Computer Interaction System 037

Task 2　Cooperation Between Eyes and Legs........................... 042

Project 2　Attitude Control of Micromouse 044

Task 1　Micromouse Runs 044

Task 2　Micromouse Turns 052

Chapter 3　Advanced Skills and Competitions 057

Project 1　Analysis of Common Algorithms............................... 059

Task 1　The Left-Hand Rule 059

Task 2　The Right-Hand Rule 061

Project 2　Advanced Control Function of Micromouse................... 063

Task 1　Principle and Implementation of Multi-Sensor Cooperation 063

Task 2　Principle and Implementation of Multithreading 068

Task 3　Micromouse Avoids Obstacles and Runs Flexibly 071

Chapter 4　Extended Application 079

Project 1　Structural Composition of TQD-IOT Engineering Innovation

Course Platform 081

Task 1　Relationship Between Engineering Innovation

Course Platform and Micromouse 081

Task 2　Hardware Composition of Engineering Innovation

Course Platform 084

Project 2　IOT Extended Application of Micromouse Technology 088

Task 1　Micromouse Controller Controls the Lighting System 088

Task 2　Micromouse Controller Controls the Security System 097

Task 3　Micromouse Controller Controls the Display System 109

Appendix 119

Appendix A　Micromouse Competition Going Popular in the World 121

Appendix B　Entry-Level Micromouse Competition Analysis 133

Appendix C　Device List of TQD-Micromouse-JQ 139

Appendix D　Device List of TQD-IOT Engineering Innovation Course

Platform 139

Appendix E　Teaching Content and Class Arrangement 140

Appendix F　The Circuit Diagram Symbol Comparison Table 140

Appendix G　Bilingual Comparison Table of Glossary 141

Appendix H　The International Curriculum Standard for "Micromouse

Design Principles and Production Process" 143

Chapter 1

Elementary Knowledge

The Micromouse competitions have enjoyed worldwide popularity for over four decades. Micromouse is required to search the entire maze without human manipulation to find the destination. And then, Micromouse needs to select, among the many possible paths, the optimal path to reach the destination and spurt from the start to the destination as quickly as possible. Contestants are ranked by the search time plus the spurt time of Micromouse. Mazes used in competitions comply with the international standards set by the Institute of Electrical and Electronics Engineers (IEEE). In this chapter, you will gain a systematic understanding of Micromouse technology from international standard mazes of IEEE, hardware systems and software development environment. You will also learn in more detail the fundamental principles and practical operations of Micromouse.

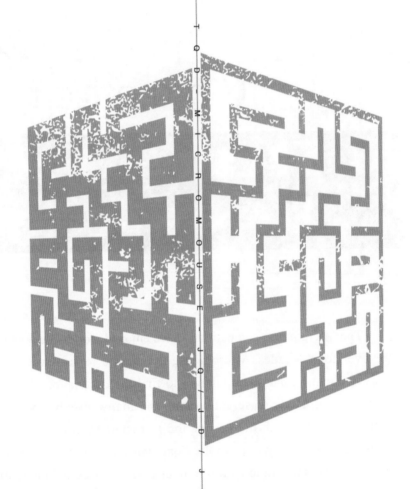

Project 1

Evolution of Micromouse

Learning objectives

(1) Learning about the evolution of Micromouse.

(2) Understanding the Micromouse competition platform, i.e. mazes and automatic scoring system.

Task 1　Origins of Micromouse

1. Birth of Micromouse

In 1938, Claude Elwood Shannon, an American mathematician born in Michigan state, completed his master's thesis *A Symbolic Analysis of Relay and Switching Circuits*. He used the Boolean algebra that happens to correspond with the binary system of 0 and 1 to process the relay switches of information in a pulse mode. The notable work had transformed the design of digital circuits both theoretically and technologically, making it an epoch-making thesis in the modern history of digital computers.

In 1948, Shannon published another famous work that is still relevant today, *A Mathematical Theory of Communication*, which earned him the title "Father of Information Theory".

In 1956, Shannon attended the Dartmouth Conference and became one of the founding fathers of the emerging discipline of artificial intelligence. He pioneered the application of artificial intelligence in computer chess and invented a mechanical mouse that could run through a maze autonomously, which proved that computers could improve their intelligence through learning.

2. Evolution of Micromouse in the World

In 1972, *Journal of Mechanical Design* started a contest where mechanical mouse solely driven by mousetrap springs competed with other entries to see which one could cover the longest distance.

In 1977, *IEEE Spectrum* introduced the concept of Micromouse, which is a small robotic vehicle controlled by microprocessors and has the capabilities to decode and navigate in complex mazes.

In 1979, the IEEE initiated a Micromouse competition through its magazine (*Spectrum and Computer*) and it rewarded the designer of the champion Micromouse that could find a way out of the maze all on its own in the shortest possible time span with USD 1,000.

In 1980, the first All Japan Micromouse International Competition was held, followed with more such events, such as UK Micromouse competition in 1980, Singapore IES Micromouse Competition in 1987, and Micromouse Competition for College Students held by China Computer Federation (CCF) in 2007. As shown in Fig.1–1–1.

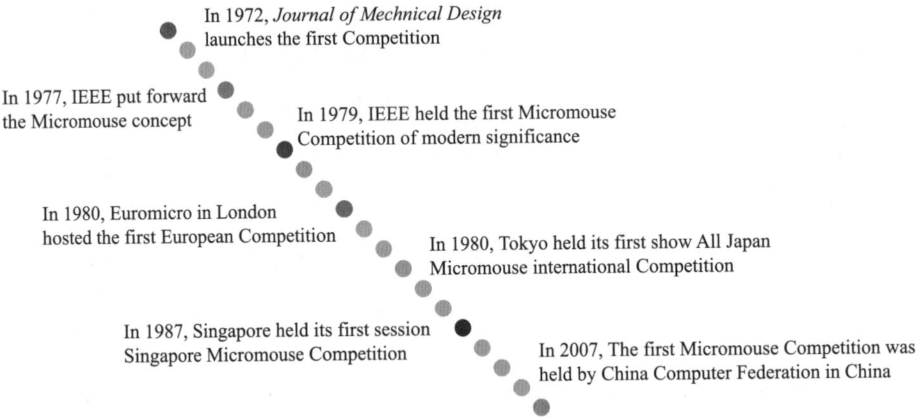

In 1972, *Journal of Mechnical Design* launches the first Competition

In 1977, IEEE put forward the Micromouse concept

In 1979, IEEE held the first Micromouse Competition of modern significance

In 1980, Euromicro in London hosted the first European Competition

In 1980, Tokyo held its first show All Japan Micromouse international Competition

In 1987, Singapore held its first session Singapore Micromouse Competition

In 2007, The first Micromouse Competition was held by China Computer Federation in China

Fig. 1–1–1 Global development trajectory

The past four decades have witnessed the great evolution of Micromouse from the mechanical mouse in 1972 to Micromouse nowadays. The competitions now feature wider participation at all education levels from around the globe. When the competitions were first launched, only graduate students from world-renowned colleges and universities such as Harvard and MIT were able to participate. Later on, students from research universities, universities of applied sciences and vocational schools could compete as well. And nowadays even primary and middle school students may take part in Micromouse competitions. Micromouse has been adopted as a teaching vehicle at educational institutions of various levels to cultivate students' engineering literacy, improve the awareness of innovation and boost their design skills.

Micromouse competitions in various forms are flourishing across the world. Now they have grown into global innovation events that are applicable to students at different education levels.

3. Evolution of Micromouse in China

Micromouse competitions have experienced over ten years of growth since 2007 in China, as shown in Fig.1–1–2. In 2007, Tianjin Qicheng Science and Technology Co., Ltd. first introduced the competition to Tianjin and added local features to create an updated version with the advanced Engineering Practice Innovation Program model as a core vision. These pioneering efforts helped boost future Micromouse events in China and played a key role in integrating relevant technologies into classroom teaching.

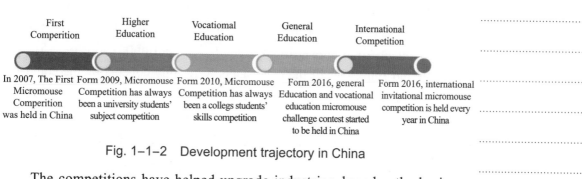

Fig. 1–1–2　Development trajectory in China

The competitions have helped upgrade industries, broaden the horizon, accumulate experience in practice and innovation and cultivate high-caliber, high-tech and highly skilled personnel (see Fig.1–1–3). A variety of Micromouse competitions has been held in China, such as contests for university students, competitions for vocational colleges, and international challenge competitions for general-vocational high schools, which enable us to gather rich experience and solid technical prowess.

The past decade has witnessed China constantly exploring new ways to make the local Micromouse competitions gain increasing international exposure.

Video

Evolution of Micromouse in China

When the competitions were first introduced into China, we simply copied foreign models, but as we gain more experience and build novel platforms for international exchanges and cooperation, foreign countries are also learning from us. Generally speaking, there are three stages in the development of Micromouse in China: imitation and learning; innovation and growth; and going global as a leader.

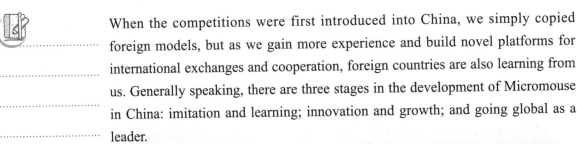

Fig. 1–1–3　Students at Micromouse competitions

The first stage: In 2015, the Tianjin team went to the United States to participate in the 30th APEC Micromouse Contest and ranked sixth globally(see Fig.1–1–4). In 2017 and 2018, Tianjin Qicheng Technology sponsored the champion team that won the enterprise designated-topic session at Tianjin College Students Micromouse Competition to go to Tokyo and compete in the 38th and 39th All Japan Micromouse International Competition(see Fig.1–1–5). The travel and boarding fees of the contestants were fully covered by the company. The two competitions in Japan enabled us to learn more about global advanced technologies of Micromouse and make connections with industry experts and leaders, greatly boosting the development of Micromouse technologies in China.

● Video

Imitation and learning

Fig. 1–1–4　Tianjin team in the US for APEC 30th Annual Micromouse Contest

Fig. 1-1-5　Tianjin team at All Japan Micromouse International Competition

The second stage: Micromouse competitions were added local, innovative features and underwent necessary reforms to comply with China's realities. A wide range of tiered teaching platforms based upon the TQD-Micromouse produced by Tianjin Qicheng Technology were created to meet the needs of students at different levels: junior high, senior high, undergraduate, and postgraduate. Since 2016, IEEE Micromouse International Invitational Competition in China has featured more extensive participation by world-famous scholars and experts, domestic and foreign teachers and students, and elite teams in China. The name list includes Professor David Otten from MIT, Professor Su Jinghui from Lunghwa University of Science and Technology in Taiwan, China, Professor Huang Mingji from Ngee Ann Polytechnic in Singapore, Professor Peter Harrison from Birmingham City University, and Mr. Yoko Nakagawa, Secretary-General of the Organizing Committee of All Japan Micromouse International Competition; faculty and students from Luban Workshop in Thailand, India, Indonesia, Pakistan and Cambodia; and competition teams from Tianjin, Beijing, Henan and Hebei, to name a few(see Fig.1-1-6 and Fig.1-1-7). By signing up for the competitions held in China, international contestants learned more about the Chinese standards, rules, models and philosophy and later accepted them. In this way, global exchanges and collaboration was facilitated and both sides had something meaningful to learn from each other.

The third stage: Educational opening-up is integral to China's reform and opening-up initiative. As "One Belt and One Road" initiative gains momentum, the Luban Workshop programme has been launched since 2016 under the guidance of the Ministry of Education. Micromouse, an exemplar of China's excellent teaching aid, has gone global thanks to the programme. Since then,

Video

Innovation and growth

• Video

Going global
as a leader

Tianjin Qicheng Technology has gone to a raft of foreign countries such as Thailand, India, Indonesia, Pakistan, Cambodia, Nigeria and Egypt to promote Micromouse competitions and offer training sessions free of charge, which are well-received by both the local teachers and students (see Fig.1–1–8-Fig.1–1–13). Micromouse has served as a bridge connecting China with the rest of the world!

Fig. 1–1–6　2018 Third IEEE Micromouse International Invitation

Fig. 1–1–7　2019 "Qicheng Cup" IEEE Micromouse International Invitation

Fig. 1–1–8　Micromouse training session at Luban Workshop in India

Fig. 1-1-9　Micromouse training session at Luban Workshop in Thailand in 2016

Fig. 1-1-10　Micromouse training session at Luban Workshop in Indonesia in 2017

Fig. 1-1-11　Micromouse training session at Luban Workshop in Pakistan in 2018

Fig. 1-1-12　Micromouse training session at Luban Workshop in Cambodia in 2018

Fig. 1–1–13　Micromouse training session at Luban Workshop in Egypt in 2020

Task 2　Competition and Debugging Environment of Micromouse

1. Competition maze

At present, the international Micromouse competition adopts IEEE standard and uses the same specification maze, that is, a square maze composed of 8×8 cells.The "walls" of the maze can be inserted, so that a variety of mazes can be formed.

The TQD-Micromouse Maze 8×8 is shown in Fig.1–1–14. The floor of the maze is 2.96 m×2.96 m, and there are 8×8 standard maze cells on it. Wall and post of the classical Micromouse maze is shown as Fig.1–1–15.

Fig. 1–1–14　TQD-Micromouse
Maze 8×8

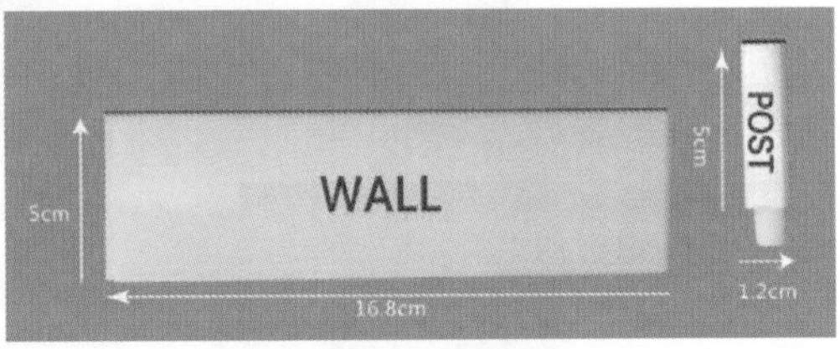

Fig. 1–1–15　Wall and post of the classical
Micromouse maze

TQD-Micromouse maze 8×8 specifications are as follows:

(1) The maze is composed of 8×8 square cells with the size of 18 cm×18 cm.

(2) The height of the walls are 5 cm and their thickness are 1.2 cm, so the actual distance of the passageways are 16.8 cm, and the walls seal the whole maze.

(3) The side of the walls are white, and the top are red. The floor of the maze is painted in black. It is made of wood, finished with non-gloss black paint. The paint on the side and top of the wall can reflect infrared light, and the floor can absorb infrared light.

(4) The start can be set at one of the four corners. The start must have three walls and only one exit. The destination is located at the center of the maze, which is composed of four cells.

(5) There are small posts, each 1.2 cm×1.2 cm×5 cm, can be inserted at the four corners of each cell. The position of the posts are called lattice points. There are at least one wall to a lattice point except for the destination.

(6) The dimensional accuracy error of the maze making should be no larger than 5%, or less than 2 cm. The joint of the maze floors shall not be more than 0.5 cm, and the gradient change of the joint point shall not be more than 4°. The gap between the walls and posts shall not be more than 1 mm.

(7) The start and the destination shall be designed based on IEEE Micromouse competition rules and standards, that is, Micromouse starts in a clockwise direction.

2. Special testing site

There are 13 marked positions painted on the special testing site and different colors are used to distinguish them (see Fig.1-1-16). They are used for aiding adjusting infrared and turning parameters. Next, let's to learn them:

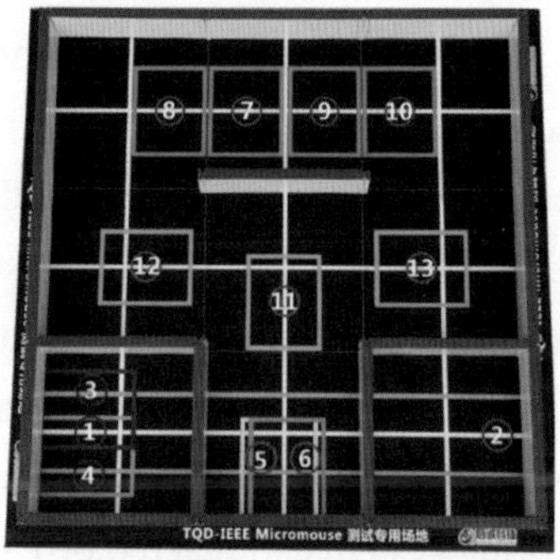

(1) ① to ②, gray passageways, which is used to detect the offset of Micromouse in the absence of infrared calibration.

(2) ③ dark red rectangle, ④ orange rectangle; ③ to ②, ④ to ② are both used to check Micromouse's forward going condition with infrared calibration.

Fig. 1-1-16   Special testing site for TQD-IEEE Micromouse

(3) ⑤ yellow rectangle is used to adjust the left front infrared intensity of Micromouse, ⑥ green rectangle is used to adjust the right front infrared intensity of Micromouse ; Correct the attitude.

(4) ⑦, ⑧ green rectangles are used to adjust the right rear infrared intensity of Micromouse , ⑨, ⑩ green rectangles are used to adjust the left rear infrared intensity of Micromouse ; Detect the intersection.

(5) ⑪、 ⑫、 ⑬three blue rectangles are used to debugging 90-degree turning of Micromouse.

3. Automatic scoring system

In order to accurately measure the time used by Micromouse to complete

• Video

The working
principle of
automatic
scoring system

the competition, it is necessary to calculate the time of Micromouse passing the start and the destination full-automatically. The electronic automatic scoring system designed and produced by Tianjin Qicheng Science and Technology Co., Ltd., which is specially used for Micromouse competition, is shown in Fig.1–1–17.

TQD-Micromouse Timer V2.0 system includes the start infrared detection module, the destination infrared detection module, the scoring system module, and the scoring software, etc.

The start infrared detection module and the destination infrared detection module are charged through mini USB. Through a set of inner-placed thru-laser sensors, it can detect the passing of Micromouse. The scoring system module is used to receive the data sent by the infrared detection modules through ZigBee. After been processed by the scoring software in computer, the running condition of Micromouse in the maze is shown in a visualized manner. The scoring software can also be used alone. The start event and destination event can be input through mouse. The overall timing accuracy of the scoring system can reach 0.001 s.

Fig. 1–1–17 TQD-Micromouse Timer V2.0

The start infrared detection module and the destination infrared detection module are respectively installed in the start cell and the destination cell, as shown in Fig.1–1–18, Fig.1–1–19. When Micromouse passes by, laser beam is blocked, thus generates a start or destination signal.

Fig.1–1–18 The start

Fig.1–1–19 The destination

Reflection and Summary

(1) What are the components of the Micromouse maze based on IEEE standard?

(2) What are the characteristics of Micromouse competition?

(3) The usage automatic scoring system has improved the accuracy of competition result calculation greatly. Please briefly explain it's working principle.

Project 2

Micromouse Hardware Structure

 Learning objectives

(1) Understanding the basic hardware structure of Micromouse.

(2) Learning how Micromouse hardware works.

This book uses TQD-Micromouse-JQ as the teaching carrier, as shown in Fig. 1–2–1. TQD-Micromouse-JQ uses Arduino, the mainstream of

Fig. 1–2–1　TQD-Micromouse-JQ

the international maker, as the core processor, provides a graphical software programming environment, and enhances the learning interest of beginners; it is suitable for the practical teaching of entry-level students for Micromouse competition, and is the preferred platform for cultivating students' engineering literacy and scientific and technological innovation ability.

Task 1　Components of Micromouse

1. Circuit composition of TQD-Micromouse-JQ

TQD-Micromouse-JQ mainly has the following structural characteristics:

(1) The main control chip adopts AVR ATmega328P with high performance and low power consumption;

(2) Built-in HC06 bluetooth module, real-time high-speed and accurate data transmission, the effective distance can reach more than 10 meters;

(3) APP online debugging is flexible and convenient, and realizes the function of Micromouse and mobile bluetooth wireless communication;

(4) Five sets of high-precision infrared digital sensors to detect the information of the labyrinth baffle in all directions;

(5) Adopt steel gear precision N20 geared motor, working voltage is 3-6 V, strong anti-interference ability;

(6) The exquisite Micromouse body shell adopts 3D printing DIY design, free splicing and unlimited creativity;

(7) Easy-to-learn, easy-to-understand, and easy-to-practice graphical programming based on the international open source Arduino, making learning vivid and interesting;

(8) TQD special DEMO program package, including on-site debugging, optimization algorithm and other programs, allowing zero-based learners to quickly get started and improve their ability quickly.

The circuit composition of TQD-Micromouse-JQ is shown in Fig. 1–2–2, which mainly includes six parts: core control circuit, power supply circuit, infrared detection circuit, Bluetooth circuit, motor drive circuit and digital/analog expansion interface. The most important part is the core control circuit, and other components depend on it to work. The power supply circuit drives the core control circuit, which then interacts with several other parts to realize the function of Micromouse.

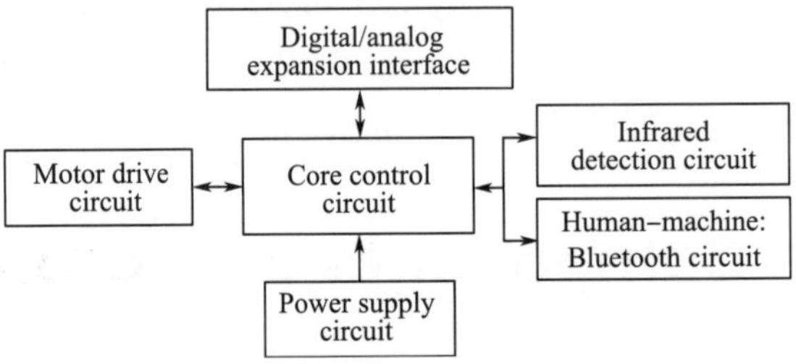

Fig. 1–2–2　The circuit composition of TQD-Micromouse-JQ

2. The frame structure of TQD-Micromouse-JQ

Before reading this book, readers may not have any concept of Micromouse, but if it comes to robots, everyone is no stranger. So how is a robot defined? It can be said that a robot is an automated machine that can achieve certain or certain special tasks. This kind of machine has certain intelligence, such as certain perception, planning, action, and coordination capabilities. In this sense, the essence of Micromouse is a kind of miniature intelligent mobile robot.

Although robots have different intelligence and tasks to achieve, their structure can be roughly divided into several parts, namely the sensor part, the main control chip part, and the actuator part.

The component layout of the TQD-Micromouse-JQ is shown in Fig. 1–2–3.

1) Sensor part

Micromouse have a certain degree of intelligence. In order to be able to respond to the external environment, they need to obtain external information, including sound, light, electricity, magnetism, temperature and humidity, obstacle information in the environment, and so on. Sensors used to obtain external information are like human eyes and ears.

2) Main control chip part

This is the core component of Micromouse. It receives the signal from the sensor part, and according to the decision system (software program)

Video

Compositions of TQD-Micromouse-JQ

Fig. 1–2–3　The component layout of the TQD-Micromouse-JQ

written in advance, determines the response to the external signal and sends the control signal to the actuator part. Its function is like the human brain.

3) Actuator part

Micromouse complete different behaviors or actions through actuators, such as lighting up light-emitting diodes and making sounds, but for Micromouse, the most basic actuator is tires, which are like human limbs.

Task 2　The Control Structure of Micromouse

AVR microcontroller is a relatively novel microcontroller introduced by Atmel. Its notable features are high performance, high speed, low power consumption, built-in Flash RISC (reduced instruction set computer). The microcontroller has the following characteristics:

(1) AVR microcontroller instructions are in units of words, and most instructions are single-cycle instructions. The next instruction is read while executing the instruction. Usually the clock frequency is 4-8 MHz, so the shortest instruction execution time is 250-125 ns.

(2) The I/O port resources of the AVR microcontroller are flexible and powerful. All I/O lines have configurable pull-up resistors, which can be individually set as input/output, and can be set (initial) high-impedance input. Strong driving capability (power driving devices can be omitted). Equipped with multiple independent clock dividers for UART, IIC, and SPI. Among them, the prescaler with up to 10 bits is matched with 8 to 16-bit timers, and the frequency division coefficient can be set by software to provide multiple levels of timing time.

At present, AVR has been widely used in various fields such as computer peripheral equipment, industrial real-time control, instrumentation, communication equipment and household appliances.

The core board is the brain center of Micromouse. The AVR microcontroller ATmega328P is shown in Fig. 1–2–4.

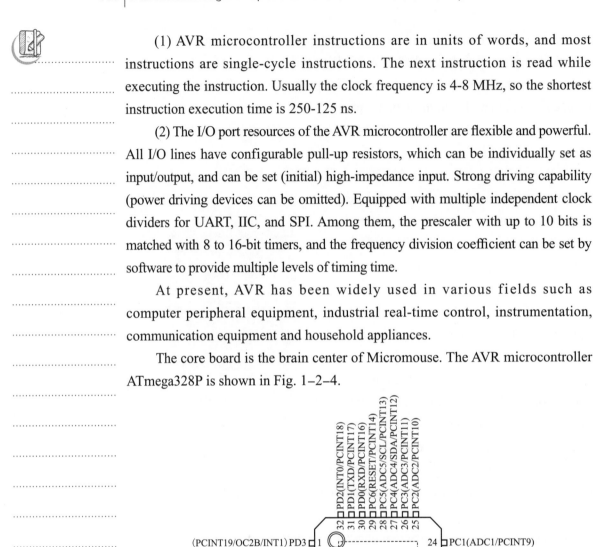

Fig. 1–2–4　ATmega328P chip pin diagram

1) Main features of the core controller

Processor: Atmel Atmega328P.

Digital I/O (digital input/output port): D0-D13.

Analog I/O (analog input/output port): A0-A5.

Input voltage: USB interface power supply or 5-12V external power supply.

Output voltage: support DC 3.3 V/5 V output.

Among them, D3, D5, D6, D9, D10, D11 ports in Digital I/O can double as PWM output interfaces.

2) Support mobile phone online debugging

The control board integrates the bluetooth module HC-06 (see Fig.1–2–5), and the communication distance is up to 10 m.

Fig. 1–2–5   Bluetooth module HC-06

The small red light of the bluetooth module flashes, indicating that the bluetooth wireless connection has not been established; if it is on, it indicates that the bluetooth wireless connection has been established.

Reflection and Summary

(1) How to transfer data between the various modules of Micromouse?

(2) Are there similar electronic components for the three major parts of Micromouse?

(3) TQD-Micromouse-JQ can support 12 digital signals and 6 analog signals at the same time. Simple setting of I/O pins can realize signal reading and output.

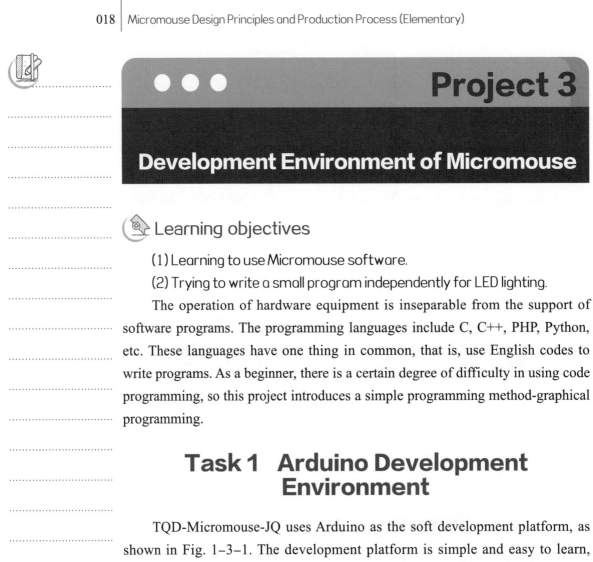

Project 3

Development Environment of Micromouse

Learning objectives

(1) Learning to use Micromouse software.

(2) Trying to write a small program independently for LED lighting.

The operation of hardware equipment is inseparable from the support of software programs. The programming languages include C, C++, PHP, Python, etc. These languages have one thing in common, that is, use English codes to write programs. As a beginner, there is a certain degree of difficulty in using code programming, so this project introduces a simple programming method-graphical programming.

Task 1　Arduino Development Environment

TQD-Micromouse-JQ uses Arduino as the soft development platform, as shown in Fig. 1–3–1. The development platform is simple and easy to learn, highly interactive and creative. The programming language is extremely easy to master, and at the same time it has sufficient flexibility. It does not require a lot of microcontroller foundation and programming foundation. After simple learning, it can be developed quickly.

The Arduino in this book uses version 1.6.4 as an example. The installation method is very similar to the usual installation. You don't need to make any changes. Just click the "Next" button until the installation is complete.

The software after the initial installation can only enter codes, readers can enter codes to write programs.

For junior learners, another programming method is also provided, that is, graphical programming, which is similar to building blocks. This method

● Software

Downloading
Arduino

greatly reduces the difficulty of programming. Even electronics enthusiasts without programming experience, including teenagers and even children, can try to implement their own designs for the Arduino controller in a graphical and interesting programming method.

Next, let's add graphical programming function to the Arduino software:

(1) Select the "File" → "Preferences" command, as shown in Fig. 1-3-2 and Fig. 1-3-3.

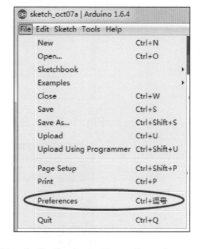

Fig. 1-3-1 Arduino software development platform

Fig. 1-3-2 Location of preferences

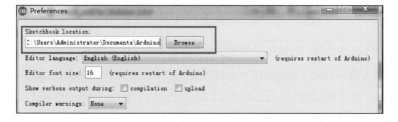

Fig. 1-3-3 Project folder location

(2) Copy the Jar package. Copy the tools folder (in the "Ardublock Jar package" folder shown in Fig. 1-3-4) to the project folder location as shown in Fig. 1-3-5.

Fig. 1-3-4 Jar package

Fig. 1-3-5 Project folder location

So far, the graphical programming function has been added, as shown in Fig. 1–3–6. Restart the software, you can see an additional "ArduBlock" under the "Tools" column, click to enter the graphical programming interface, as shown in Fig. 1–3–7.

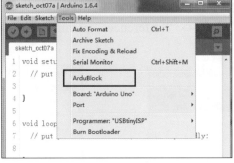

Fig. 1–3–6　Successfully added ArduBlock

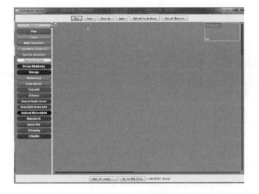

Fig. 1–3–7　Graphical programming interface

(3) LCD, voice recognition and timing interruption, Scoop support library added. Find the installation location of Arduino in your computer, and completely delete the LiquidCrystal folder under Arduino/libraries. Then copy the three folders of LiquidCrystal, voiceRecognition and MsTimer2 to the original location (do not overwrite, otherwise you need to reinstall Arduino), as shown in Fig. 1–3–8.

Fig. 1–3–8　LCD liquid crystal, voice recognition and timer interrupt support library

So far, the graphical programming function has been added. Please restart the software to start calling.

Task 2　Logical Thinking of Micromouse

Write the first program in Arduino—to light up a small single-color LED.

1. Graphical programming interface

Open the graphical programming interface, you will see a column of columns with different colors on the left. The first eight columns that will be used in subsequent experiments are also shown in Fig. 1–3–9. Click the button to view the columns details.

Fig. 1–3–9　The first eight columns

Ardublock graphical programming, designs graphics of different colors and shapes for different programming statements, and only graphics that conform to the programming syntax can be connected together. The following is an introduction to these eight columns.

Video

Introduction to common toolbars of Arduino

1) "Control" column

It contains statements that control the operation of the program, such as the main program, conditional judgment, loop, and delay.

The operation of the program first needs a main function, as shown in Fig. 1–3–10. The difference between the two main function modules in Fig. 1–3–10 is that the former has one more setting, which can set the initial values of some variables.

Similarly, the difference between these two conditional judgment function modules (see Fig. 1–3–11) is whether there is content that needs to be executed when the condition is not met.

Fig. 1-3-10　Main function module

Fig. 1-3-11　Conditional judgment function module

2) "Pin" column

It contains the digital pins and analog pins corresponding to sensors and actuators.

It can be seen that according to the different data types, pins are divided into two categories:

Fig. 1-3-12　Digital and analog pins

digital pins and analog pins (see Fig. 1-3-12). Each module is divided into small and large. The small module can only choose the pin number, which is used to indicate the sensor, such as buttons, temperature, light, etc. The large module can also set the high and low level or analog quantity, which is used to indicate the actuator, such as LED, fan, buzzer, etc.

3) "Tests" column

It contains graphics related to logical judgments. Different data types require different graphics.

Integer type—integers such as 0, 1, 2, 3;

Character type—'a', 'b', 'c' and 'd' etc.;

Digital type—high level, low level.

Different data types have their own shapes, this must be paid attention to when using!

4) "Mathematics" column

It contains mathematical operations such as addition, subtraction, multiplication and division. Of course, all that can be used for mathematical operations must be integers or integer variables.

5) "Variable/Constant" column

It contains integer, character, numeric variable settings, constant settings, etc., as shown in Fig. 1-3-13, from top to bottom are integer variables, character

Text

Digital signal and analog signal

variables, and numeric variables. Commonly used, such as the distance measured by ultrasound and the size of the temperature can be set as integer variables; the wireless communication between the mobile phone APP and the experimental platform can be set as character variables; and the state of the buttons and the state of the infrared sensor need to be set as numeric variable.

6) "Generic Hardware" column

It contains several general hardware, such as LCD module, steering gear, ultrasonic, etc.

As shown in Fig. 1–3–14, the steering gear and ultrasonic are very simple to use, just set the correct angle and pin number.

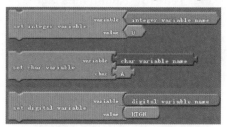

Fig. 1–3–13   Variable module

Fig. 1–3–14   SG90 steering gear and ultrasonic module

The LCD module must be 1602 below (the same model as the experimental platform), and the wrong choice will not be used.

It is important to note that when the data refresh is relatively fast, you also need to add a delay function and a clear screen function to assist in reading the data, as shown in Fig. 1–3–15.

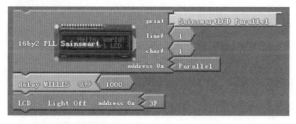

Fig. 1–3–15   LCD module

7) "Communication" column

It contains graphics such as reading serial port, serial printing, etc. as shown in Fig. 1–3–16.

Fig. 1–3–16　Serial printing

The reading serial port graphic is used to read wireless communication data, as shown in Fig. 1–3–17. Serial printing graphics is used to return information, as shown in Fig. 1–3–18.

Fig. 1–3–17　Read serial port data　　　　Fig. 1–3–18　Return message

8) "Scoop" column

The main modules used in the "Scoop" column include the main program and the delay module, as shown in Fig.1–3–19, Fig.1–3–20.

Fig. 1–3–19　Main function module　　　　Fig. 1–3–20　Delay module

Understanding the functions of each column will be of great help to future experiments.

Next, try to write a program to light up the single-color LED module. The purpose of the experiment is to input a high level to the LED to make it glow.

2. Flow chart

The LED lighting flow chart is shown in Fig. 1–3–21.

3. Graphical programming

(1) First, the program runs from the main function, so the "Main Program" module in the "Control" column is needed, as shown in Fig. 1–3–22.

Video

Lighting LED

 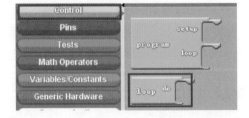

Fig. 1–3–21　LED lighting flow chart　　　Fig. 1–3–22　"Main Program" module

(2) Lighting up the LED is to output a high-level signal to the LED, and the "Set Digital Pin Value" module in the "Pin" column is needed, as shown in Fig. 1–3–23.

(3) Combine these two modules, and connect the single-color LED module to the Micromouse core controller, such as pin 8. Set the correct pin number and high and low level for the graphical programming module, as shown in Fig. 1–3–24.

Fig. 1–3–23　"Set digital pin value"　Fig. 1–3–24　Set the correct pin number
module　　　　　　　　　　　　and high and low level

At this point, the program for lighting the LED is completed. Next, connect the core controller to the computer, select the corresponding serial port number, download the program, and observe whether the LED is on, as shown in Fig. 1–3–25.

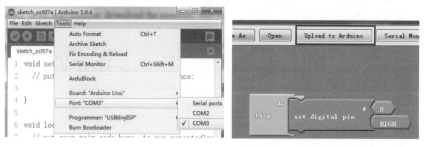

Fig. 1–3–25　Turn on the LED

Reflection and Summary

(1) What is the difference between digital signal and analog signal?

(2) What kind of signal does the light intensity and the key signal belong to?

(3) Graphical programming is encapsulated for code programming, and different graphics correspond to different programming statements; shapes and colors are different, and only graphics that conform to the programming rules can be connected together, which greatly reduces the programming difficulty for beginners.

Project 4

Basic Function Control of Micromouse

Learning objectives

(1) Learning how to debug the Micromouse infrared.

(2) Learning how to drive the Micromouse motor.

The basic function of Micromouse is to run from the start to the destination. Fast and accurate walking in a maze of limited width and a large number of turns are inseparable from high-precision sensor detection and motor operation control. In different environments, the light intensity is different, and the ground friction also has certain differences. Therefore, the human-computer interaction method must be used to debug the infrared detection accuracy of Micromouse and the motor speed.

Task 1   Micromouse Sees the World

1. The role of robot sensors

Sensors play a very important role in the control of robots. Because of sensors, robots have the perceptual function and response ability similar to humans. The perception system is an essential part of robots to achieve autonomy, and sensors are an essential part of the perception system.

According to the function of sensors, they can be divided into internal sensors and external sensors. Internal sensors mainly measure the internal systems of the robot, such as temperature, motor speed, motor load, battery voltage, etc.; external sensors mainly measure the external environment, such as distance, sound, light, etc. According to the operating mode of the sensor, the sensors used in the mobile robot are divided into passive sensors and active sensors. Passive sensors themselves do not emit energy, such as CCD and CMOS camera sensors;

active sensors emit detection signals, such as ultrasonic, infrared, and laser. However, the reflected signals of this type will be affected by many substances, which will affect the acquisition of accurate signals. At the same time, signals are also susceptible to interference.

The following describes how to detect the labyrinth baffle through the infrared sensor. The infrared sensor is detected by the principle of triangulation. The triangulation method is to install the transmitter and receiver at a certain angle to form a triangle with the detected point. Since the distance between the transmitter and the receiver is known, the launch angle is known that the reflection angle can also be detected, so the distance from the detection point to the transmitter can be calculated. This measurement method can measure very close objects, the most accurate at present can reach a resolution of 1μm, but it cannot detect distant objects.

TQD-Micromouse-JQ is equipped with five sets of high-precision digital infrared sensors, as shown in Fig. 1–4–1. When the receiver receives IR light reflected by the obstacle, the information of the wall around Micromouse can be obtained. Based on these walls informations, the controller combines logic algorithms to control Micromouse to walk intelligently.

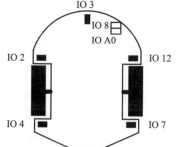

Fig. 1–4–1　The bottom board diagram of Micromouse

2. Infrared circuit composition of Micromouse

The infrared detection circuits of Micromouse are equivalent to the "eyes" of Micromouse. They are used as the input module to detect the maze walls. They are divided into five groups: left front, front, right front, left rear, and right rear. It can be seen from the above that whether the infrared emission intensity are accurate are the key to whether Micromouse can walk automatically and accurately.

The five transmitters are all connected to a common transmitting pin, IO_11, to send out a specific PWM wave.

The five groups of infrared sensors have the same circuit principle, as shown in Fig. 1–4–2.

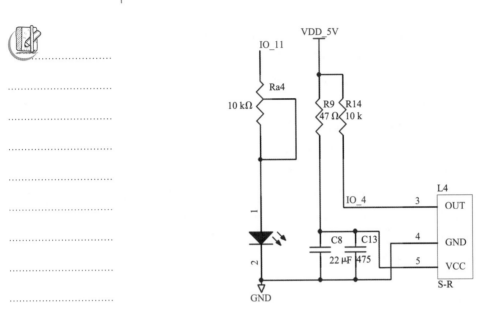

Fig. 1–4–2　Circuit diagram of infrared sensor[1]

The receiver is an integrated infrared receiving sensor whose model is IRM8601S. The receiver integrates an automatic gain control circuit, a band-pass filter circuit, a decoding circuit and an output drive circuit. The receiver is most sensitive to infrared signals with a carrier frequency of 38 kHz. When it detects an effective infrared signal, it outputs a low level, otherwise it outputs a high level. Ra4 is a current-limiting adjustable resistor, used to adjust the intensity of infrared radiation.

3. Working principle of infrared sensor IRM8601S

IRM8601S is an infrared receiver with a working voltage of 5 V and a receiving distance of 8 m. The appearance is shown in Fig. 1–4–3, and the three pins are OUTPUT, GND and VCC.

In order to facilitate the understanding of the working principle of the infrared receiver, first introduce the concept of modulation. Modulation is a process in

Fig. 1–4–3　Appearance of infrared receiver IRM8601S

1—OUTPUT；2—GND；3—VCC

Text

Extended knowledge of sensors

---

[1] Similar drawings are schematic diagrams derived from Protel 99SE, and their graphic symbols are inconsistent with the national standard symbols. Please refer to the Appendix F for the comparison between the two.

which an input signal carrying information is used to control a certain parameter of another signal so that it changes according to the law of the input signal. The input signal is called the modulation signal, the signal to be controlled is called the carrier (or carrier frequency) signal, and the output signal is the modulation wave, as shown in Fig. 1–4–4. Modulation waves are divided into amplitude modulation waves, frequency modulation waves and phase modulation waves according to the type of carrier signal parameters controlled by the modulation signal.

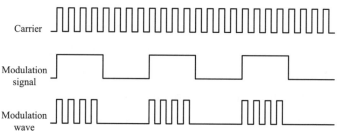

Fig. 1–4–4　Modulation diagram

The center frequency of the band-pass filter inside the IRM8601S sensor is 38 kHz, so the sensor is most sensitive when the carrier signal that drives the infrared ray is 38 kHz. According to the data manual of IRM8601S, the modulation signal should be a square wave with a period of 1 200 μs. The modulation wave driving the infrared emission is shown in Fig. 1–4–5.

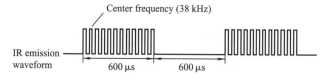

Fig. 1–4–5　Modulated wave driving infrared emission

When IRM8601S detects a signal, it outputs an effective level (low level). The range of the effective level maintenance time T_{WL} is 400 μs$<T_{WL}<$800 μs, and the sensor output waveform is shown in Fig. 1–4–6.

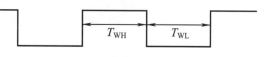

Fig. 1–4–6　Sensor output waveform

T_{WH}: High level duration in a cycle;

T_{WL}: Low level duration in a cycle.

4. Infrared emission intensity adjustment

In order for the Micromouse to walk normally and smoothly in the maze, it first needs to adjust its four infrared sensors before and after it.

Using a dedicated test site, you only need to place Micromouse in a suitable marking position to quantitatively adjust the intensity of the infrared sensor, completely solve the inaccurate debugging of the infrared sensor, and make the debugging of the infrared sensor easier and more scientific. When Micromouse is walking on the dedicated test site, it will automatically correct the vehicle's posture and detect intersections and turn according to the results of the infrared sensor detection, realizing the intelligence of Micromouse.

Before starting to debug the infrared sensor emission intensity, we write a small program. When the sensor detects the wall, it will send data to the mobile phone to tell us whether Micromouse "sees" the obstacle:

(1) In the application part of the Arduino software, we have learned how to emit a certain frequency of PWM waves (using the "tone" and "no tone" graphics).

Combined with the characteristics of the infrared receiver IRM8601S, the transmission time should be kept at about 600 μs, and the transmission must be turned off for the same time. Since the infrared sensor has the farthest detection distance when the PWM wave frequency is 38 kHz, take a frequency between 20~38 kHz, such as 28 kHz. When adjusting the emission intensity later, modify the emission frequency appropriately according to the actual situation, as shown in Fig. 1–4–7.

(2) Naming as variables for the four sensors below, as shown in Fig. 1–4–8.

Fig. 1–4–7　Infrared PWM wave drive　　　Fig. 1–4–8　Set sensor pin

L_correct: Left front sensor, corresponding to IO 2;

R_correct: Right front sensor, corresponding to IO 12;

L_turn: Left rear sensor, corresponding to IO 4;

R_turn: Right rear sensor, corresponding to IO 7.

Adding four "if " graphics to determine whether the infrared has detected walls, as shown in Fig. 1–4–9.

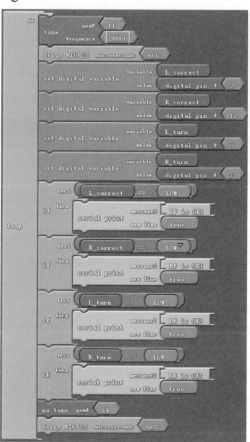

Fig. 1–4–9 Infrared test

Task 2 Micromouse Moves

1. Introduction of motor

A motor refers to an electromagnetic device that realizes the conversion or transmission of electric energy according to the law of electromagnetic induction,

Text

Extended knowledge of motion structure

and can be divided into a eletric motor and a generator. Electric motors convert electrical energy into mechanical energy, and generators convert mechanical energy into electrical energy. Motors are represented by the letter "M" in the circuit (the old standard uses "D"). Its main function is to generate driving torque as a power source for electrical appliances or various machinery. Motors can be divided into high-speed motors, low-speed motors, constant-speed motors, and speed-regulating motors according to their operating speed. Low-speed motors are divided into gear reduction motors, electromagnetic reduction motors, torque motors and claw-pole synchronous motors.

The TQD-Micromouse-JQ Micromouse uses an upgraded version of high-quality steel gear N20 geared motor (commonly known as "motor") as its power source, as shown in Fig. 1–4–10.

Fig. 1–4–10　N20 geared motor

2. Drive N20 geared motor

After the previous study, We have a more in-depth understanding of Arduino software and hardware development. Next, we can start the motor through simple debugging.

The N20 geared motor is simple to use. When the input voltage of the two pin wires is high and low, the motor can be rotated forward and reverse, as shown in Fig. 1–4–11.

The left motor occupies pins 9 and 10, and the right motor occupies pins 5 and pin 6. The rotation truth table of the two motors is shown in Table 1–4–1.

Fig. 1–4–11 N20 geared motor pin number

Table 1–4–1 The rotation truth table of the two motor

Pin	9 pin Low level	9 pin High level	Pin	6 pin Low level	6 pin High level
10-pin Low level	stop	Forward	5-pin Low level	stop	Forward
10-pin High level	Reverse	stop	5-pin High level	Reverse	stop

● Video

Experiment_
Micromouse
Runs

In the program shown in Fig. 1–4–12, input high level to pin 9 and pin 6, and input low level to pin 10 and pin 5. You can see after downloading the program, the motor rotates counterclockwise, which is customarily called forward rotation.

Fig. 1–4–12　test program

Similarly, you can reverse it, input low level to pin 9 and pin 6, and input high level to pin 10 and pin 5, then the motor will rotate clockwise, that is, reverse. Different motor modules may have different directions of forward and reverse rotation

It can not only control the forward and reverse rotation of the motor, but also control the speed of its rotation. But this requires the use of PWM technology to adjust the output analog quantity.

Regarding the two motors of the Micromouse, we are now familiar with how they operate. In other Micromouse, there are also three-wheeled, crawler, and four-legged driving methods.

3. Motor driver chip

L9110S is a semiconductor integrated product, the limit parameter is 800 mA/ 2.5~12 V, the wide power supply voltage range: 2.5~12 V. The pin diagram is shown in Fig.1–4–13.

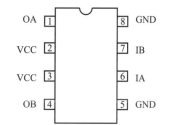

L9110S is a two-channel push-pull power amplifier dedicated integrated circuit device designed for controlling and driving motors. The

Fig. 1–4–13　The pin diagram

discrete circuit is integrated into a single IC, which reduces the cost of peripheral devices and improves the reliability of the whole machine. The chip has two TTL/CMOS compatible level inputs, which has good anti-interference; the two output terminals can directly drive the forward and reverse movement of the motor. It has a large current drive capability, and each channel can pass 800 mA continuous current, peak current capability up to 1.5 A; at the same time, it has a low output saturation voltage drop; the built-in clamping diode can release the reverse impulse current of the inductive load, making it drive relays, DC motors, stepping motors or the use of the switching power tube is safe and reliable. L9110S is widely used in toy car motor drive, pulse solenoid valve drive, stepper motor drive and switch power tube circuits, as shown in Fig. 1–4–14.

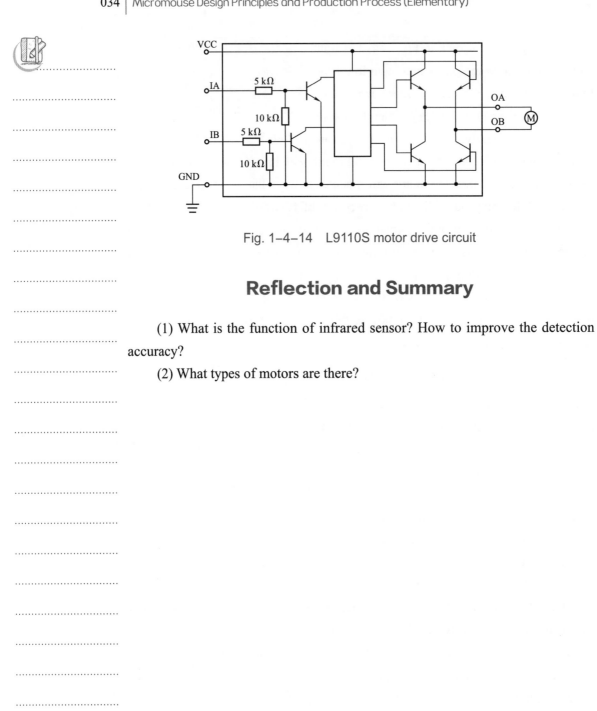

Fig. 1–4–14　L9110S motor drive circuit

Reflection and Summary

(1) What is the function of infrared sensor? How to improve the detection accuracy?

(2) What types of motors are there?

Chapter 2

Comprehensive Practice

This chapter focuses on the functional introduction of each module of Micromouse, mainly from the Micromouse human-computer interaction system, understanding the built-in Bluetooth module and the mobile terminal Bluetooth APP for real-time data transmission and information return. To know the principle of Micromouse sensor and motor working together. The attitude control, the motors drive, and the turns of Micromouse.

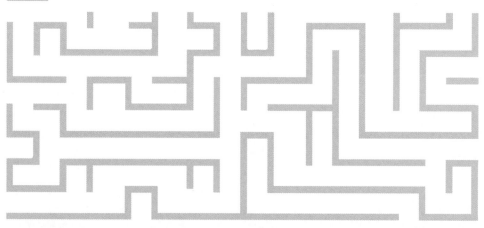

Project 1

Interaction Control of Micromouse

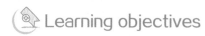 **Learning objectives**

(1) Learning how to use Bluetooth technology to control Micromouse.

(2) Learning how to use sensors to control the motors to realize the automatic control of Micromouse.

When Micromouse is running, it should have two control systems; the human-computer interaction control system is mainly the start and stop of Micromouse and the situation that needs emergency response; the automatic control system is mainly to realize the system control automatically when Micromouse is running, which can improve the response speed of the system and save the human cost.

Task 1　Human-Computer Interaction System

1. The concept of human-computer interaction

With the continuous expansion and deepening of the field of human cognition, the continuous development and improvement of the level of cognition, the space for human development continues to expand, and many tasks are unpredictable. Therefore, the development and application of human-computer interaction systems are of great urgency. The "human-computer interaction" of embedded micro-robots refers to Human-Robot, that is, the interaction between humans and robots, which uses computer input and output devices to realize human-computer dialogue in an effective manner. Human-computer interaction technology includes machines that provide people with a large amount of relevant information and prompts through output or display devices, and people use input devices to input relevant information, answer questions, and promptly ask for instructions. Human-computer interaction technology is an important part of interface design for computer users. It is closely related to cognition, ergonomics, psychology and other disciplines.

The design of human-computer interaction system should comply with the following principles: controllable principle, ease of use principle, intuitive principle, simplicity principle and visibility principle. A good robot human-computer interaction system should enable the human brain's decisions to be transmitted to the robot faster, and it should also feed back system information faster so that the user can make decisions.

2. Human-Computer interaction of TQD–Micromouse–JQ

The human-computer interaction system of Micromouse refers to the built-in Bluetooth module and the Bluetooth App on the mobile terminal for real-time data transmission, including command sending and information return.

The following is the human-computer interaction through programming: when the command "a" is sent, the message "Who are you?" is returned; when the command "b" is sent, the message "I can see the world!" is returned.

1) Bluetooth serial port software

Bluetooth is translated from the English word Bluetooth. It refers to a technical standard that enables short-distance data exchange between devices. Need to install a Bluetooth serial port software on the phone, this software can help send instructions to the development board.

There are many kinds of Bluetooth serial software, you can choose one of them to download and install. The installation method is the same as that of other Apps. The main interface of a commonly used Bluetooth serial port assistant is shown in Fig. 2–1–1.

Using the Bluetooth serial port software to connect with the built-in Bluetooth module of the core controller, select the correct device, and enter the pairing password (the default is 0000 or 1234) to connect successfully, as shown in Fig. 2–1–2.

●Software

Downloading the bluetooth App

Fig. 2–1–1　The main interface of a commonly used Bluetooth serial port assistant

Fig. 2–1–2　Bluetooth connection interface

2) Sort out logical thinking—flow chart

Now that we understand the Bluetooth human-computer interaction of Micromouse, how is it implemented?

First, you need to read the Bluetooth data and receive the instruction;

Secondly, we must determine what the instruction is;

Finally, according to different instructions, different information is output. The flow chart is shown in Fig. 2–1–3.

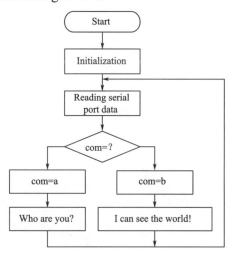

Fig. 2–1–3　Bluetooth command flow chart

3) Graphical programming

According to the previous analysis, the Bluetooth module is required to read the instructions sent by the mobile phone. When the instruction "a" is read, the message "Who are you?" is returned; when the instruction "b" is read, the message "I can see the world!" is returned. Write the program in this logical sequence below.

First, drag and drop the "Main Program" module to the programming area. Before reading the serial port, determine whether there is data in the serial port, as shown in Fig. 2–1–4.

Fig. 2–1–4　Determine whether the serial port has data

Secondly, set a character variable to save the Bluetooth command read. For ease of understanding, rename the variable name to "com", as shown in Fig. 2–1–5.

Fig. 2–1–5　Rename the variable name to "com"

Finally, judge whether the command is "a" or "b", as shown in Fig. 2–1–6, and return information, as shown in Fig. 2–1–7.

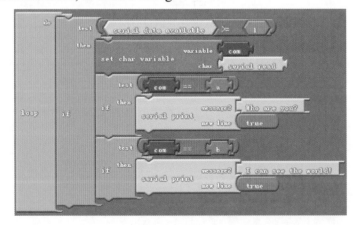

Fig. 2–1–6　Judgment instruction

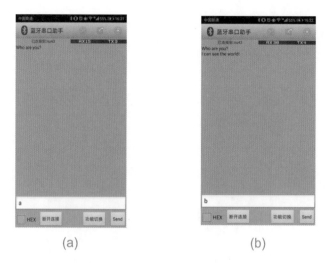

(a)   (b)

Fig. 2-1-7   Command output interface

4) Bluetooth control motor operation

According to the previous procedure, we already know how the information is transmitted between Bluetooth and the controller, then we will use Bluetooth to control the operation of the motor.

Let the pin6, the pin9 be high level, the pin5, the pin10 be low level, and both motors are forward, as shown in Fig. 2-1-8.

Let pin 5, pin 6, pin 9, and pin 10 all be low, and neither motor will rotate, as shown in Fig. 2-1-9.

Fig. 2-1-8   Motor forward   Fig. 2-1-9   Motor stop

We use the Bluetooth command "a" to control the two motors to rotate forward, and the Bluetooth command "b" to control the two motors to stop, as shown in Fig. 2-1-10.

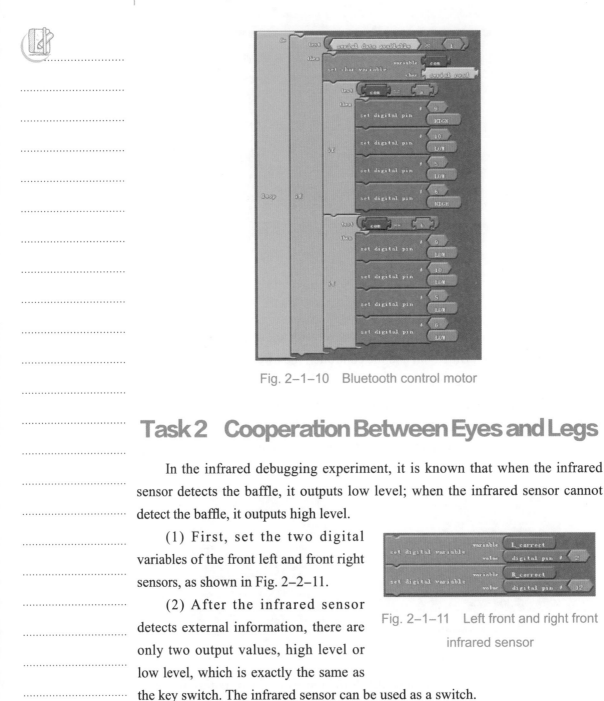

Fig. 2-1-10　Bluetooth control motor

Task 2　Cooperation Between Eyes and Legs

In the infrared debugging experiment, it is known that when the infrared sensor detects the baffle, it outputs low level; when the infrared sensor cannot detect the baffle, it outputs high level.

(1) First, set the two digital variables of the front left and front right sensors, as shown in Fig. 2-2-11.

(2) After the infrared sensor detects external information, there are only two output values, high level or low level, which is exactly the same as the key switch. The infrared sensor can be used as a switch.

Fig. 2-1-11　Left front and right front infrared sensor

(3) Set the infrared transmitter pin and set its transmit frequency.

(4) Observe the state of the level signal output by the sensor through the two operating states of the motor "forward" and "stop".

(5) Organize and combine the previous procedures, as shown in Fig. 2-2-12.

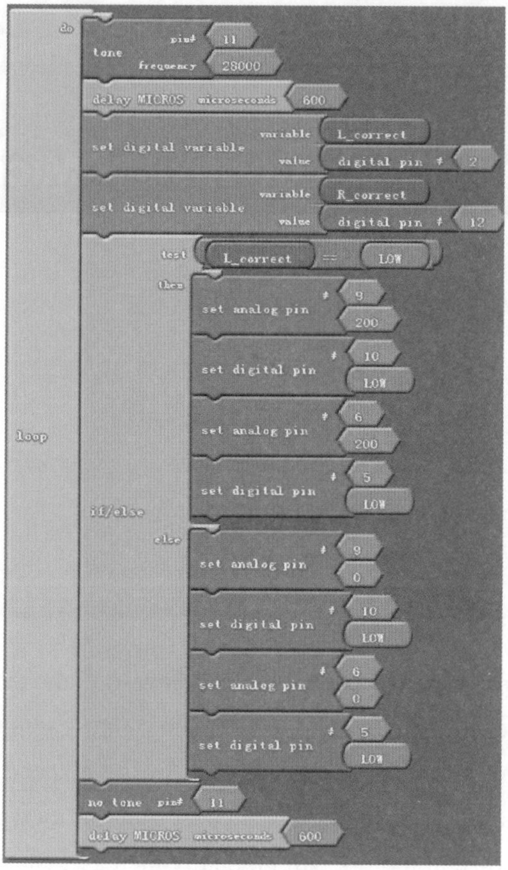

Fig. 2-1-12   Infrared sensor controls motor action.

After setting the infrared sensor emission intensity, start to download the program, place Micromouse in an open position, using objects to cover the sensors, and observe the operation of Micromouse. You can also add two sets of rear infrared sensors to the programming.

Reflection and Summary

(1) Besides Bluetooth, what are the common wireless control methods?

(2) When multiple sensors work at the same time, we need to pay attention to the interference between each other; because the transmitter uses the same pin, so the transmission frequency is the same, we can use alternate emission detection method to improve the detection accuracy.

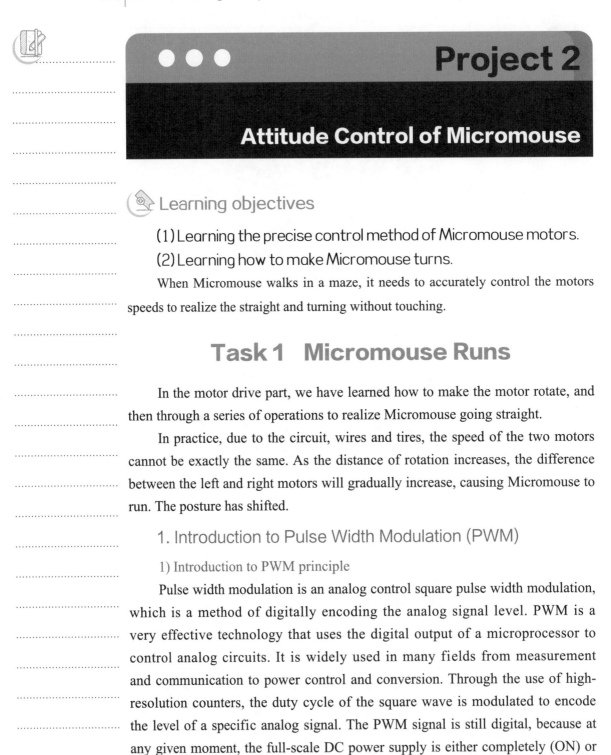

Project 2

Attitude Control of Micromouse

Learning objectives

(1) Learning the precise control method of Micromouse motors.

(2) Learning how to make Micromouse turns.

When Micromouse walks in a maze, it needs to accurately control the motors speeds to realize the straight and turning without touching.

Task 1   Micromouse Runs

In the motor drive part, we have learned how to make the motor rotate, and then through a series of operations to realize Micromouse going straight.

In practice, due to the circuit, wires and tires, the speed of the two motors cannot be exactly the same. As the distance of rotation increases, the difference between the left and right motors will gradually increase, causing Micromouse to run. The posture has shifted.

1. Introduction to Pulse Width Modulation (PWM)

1) Introduction to PWM principle

Pulse width modulation is an analog control square pulse width modulation, which is a method of digitally encoding the analog signal level. PWM is a very effective technology that uses the digital output of a microprocessor to control analog circuits. It is widely used in many fields from measurement and communication to power control and conversion. Through the use of high-resolution counters, the duty cycle of the square wave is modulated to encode the level of a specific analog signal. The PWM signal is still digital, because at any given moment, the full-scale DC power supply is either completely (ON) or completely absent (OFF). The voltage or current source is applied to the analog load in a repetitive pulse sequence of on (ON) or off (OFF). When it is on, it is when the DC power supply is added to the load, and when it is off, it is when the power supply is disconnected. As long as the bandwidth is sufficient, any analog value can be encoded using PWM.

PWM is pulse width modulation, which is a pulse waveform with variable duty cycle, as shown in Fig. 2-2-1. Control the turn-on and turn-off of the semiconductor switching device, so that a series of pulses of equal amplitude but unequal width are obtained at the output end, and these pulses are used to replace sine waves or other required waveforms. The width of each pulse is modulated according to certain rules, which can change the output voltage of the inverter circuit and the output frequency.

2) The concept of PWM and duty cycle

(1) PWM: Also known as pulse width modulation technology, it is an analog control method. The PWM waveform is shown in Fig. 2-2-2.

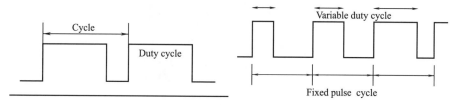

Fig. 2-2-1   Pulse width modulation   Fig. 2-2-2   PWM waveform

(2) Duty cycle: Refers to the proportion of high level in a cycle. For example, the duty cycle is 50%, that is, the high level occupies half of the entire cycle time.

3) PWM fixed frequency speed regulation principle

In the PWM speed control system, generally three methods: fixed width frequency modulation, width modulation frequency modulation, and fixed frequency width modulation can be used to change the duty cycle of the control pulse, but the first two methods change the period of the control pulse width during speed control. This causes the control pulse frequency to change. When the frequency is close to the natural frequency of the system, it will cause oscillation. In order to avoid this phenomenon, the method of changing the duty cycle with fixed frequency width modulation is used to adjust the voltage across the armature of the DC motor.

When adjusting the speed, the fixed frequency width modulation is also called the fixed frequency speed regulation. This is under the premise that the frequency of the pulse waveform does not change (the period of the pulse waveform does not change), by changing the high level time in a periodic waveform to change the duty cycle of the waveform changes the average voltage and adjusts the speed of the motor. Assuming that when the motor is always powered on, the maximum speed of the motor is v_{max} and the duty cycle is $D=t/T$, then the average speed of the motor is $v_d=D \times V_{max}$. It can be seen from the formula that when the duty cycle D is changed, we can get different average motor speed v_d, so as to achieve the purpose of speed regulation

2. Micromouse goes straight without correction

Writing a Micromouse driver program, assign high and low levels to pins 9, 10, 5, and 6 respectively, and observe the rotation direction of the left and right motors, so you can get the Micromouse straight program, as shown in Fig. 2–2–3.

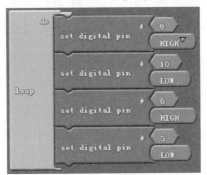

Fig. 2–2–3　Micromouse straight program

Downloading the program to Micromouse and start Micromouse to observe the deviation.

After the program is uploaded, we observe the operation of Micromouse. Does Micromouse walk in a straight line with a small offset? What should I do if the offset is large?

In fact, due to the circuit, wires and tires, the rotation speed of the two wheels cannot be exactly the same, that is to say, there must be a large or small deviation. How should it be solved?

We recall the motor speed control we did before, that is, changing the analog quantity of the motor pin can realize the control of the motor speed. So we replaced the 6 and 9 pins with analogs, and set the Bluetooth adjustment to change the size of the 6 and 9 analogs, and finally realized that Micromouse walks in a straight line or has a smaller offset within the allowable range.

(1) First, we set up two variables "l_motor" and "r_motor", and assign values to 150 respectively, as shown in Fig. 2–2–4.

(2) In order to make the program easier to understand, we introduce subroutine modules. Take out subroutines (two for each of the mother and child modules) in the "control bar", and process the Bluetooth control part and temperature control part separately. In order to ensure the correspondence between the mother and child modules of the subprogram, the names of the two must be consistent, as shown in Fig. 2–2–5.

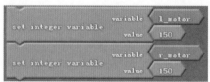

Fig. 2–2–4   Assignment procedures of left and right motor analog values respectively

Fig. 2–2–5   Left subprogram module (mother and child module)

(3) We apply subroutine modules to forward and stop respectively, as shown in Fig. 2–2–6 and Fig. 2–2–7.

In this way, only use ▢Forward▢ and ▢Stop▢ in the main program can complete the program call to start and stop Micromouse.

(4) Now we continue to add the Bluetooth control Micromouse start and stop part, as shown in Fig. 2–2–8.

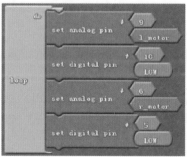

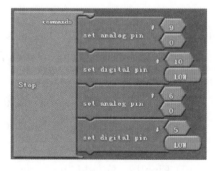

Fig. 2–2–6   Forward sub-module

Fig. 2–2–7   Stop sub-module

When the mobile phone Bluetooth sends "k", the variable "state" value is 1, and Micromouse starts to run; When the Bluetooth sends "g", the variable "state" value is 0, and Micromouse stops running.

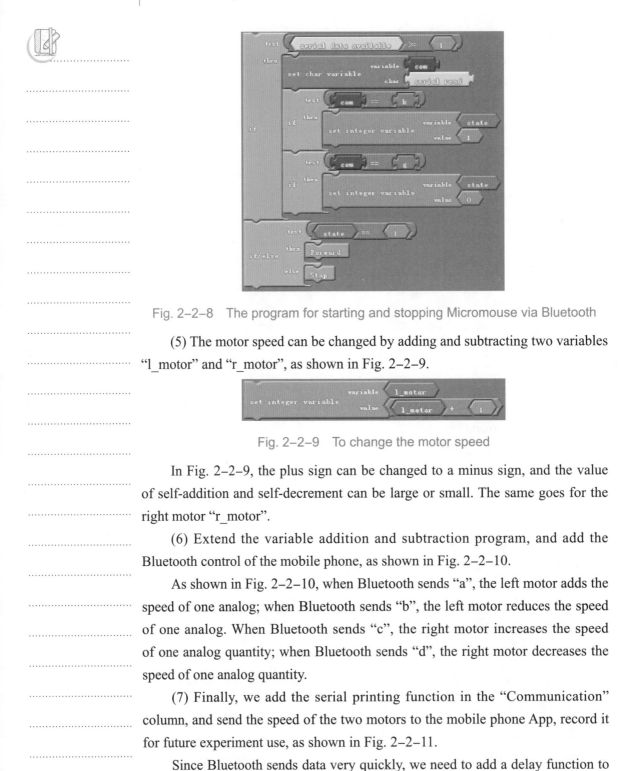

Fig. 2-2-8   The program for starting and stopping Micromouse via Bluetooth

(5) The motor speed can be changed by adding and subtracting two variables "l_motor" and "r_motor", as shown in Fig. 2-2-9.

Fig. 2-2-9   To change the motor speed

In Fig. 2-2-9, the plus sign can be changed to a minus sign, and the value of self-addition and self-decrement can be large or small. The same goes for the right motor "r_motor".

(6) Extend the variable addition and subtraction program, and add the Bluetooth control of the mobile phone, as shown in Fig. 2-2-10.

As shown in Fig. 2-2-10, when Bluetooth sends "a", the left motor adds the speed of one analog; when Bluetooth sends "b", the left motor reduces the speed of one analog. When Bluetooth sends "c", the right motor increases the speed of one analog quantity; when Bluetooth sends "d", the right motor decreases the speed of one analog quantity.

(7) Finally, we add the serial printing function in the "Communication" column, and send the speed of the two motors to the mobile phone App, record it for future experiment use, as shown in Fig. 2-2-11.

Since Bluetooth sends data very quickly, we need to add a delay function to help us get the data. We choose the "Delay" module in the "Control" column.

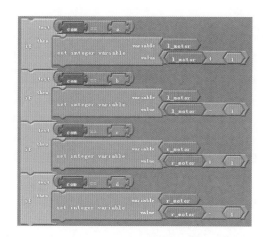

Fig. 2–2–10　Bluetooth adjustment of left and right motor speed

Fig. 2–2–11　Serial printing left and right motor speed

(8) So far, all the functions of the whole experiment have been realized. The general procedure is shown in Fig. 2–2–12.

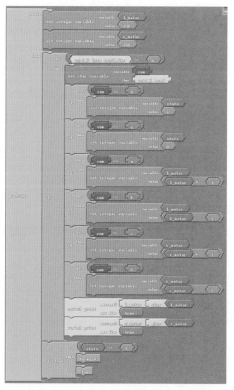

(a) Bluetooth control Micromouse to walk straight line main program module

Fig. 2–2–12　Bluetooth control Micromouse walk straight program

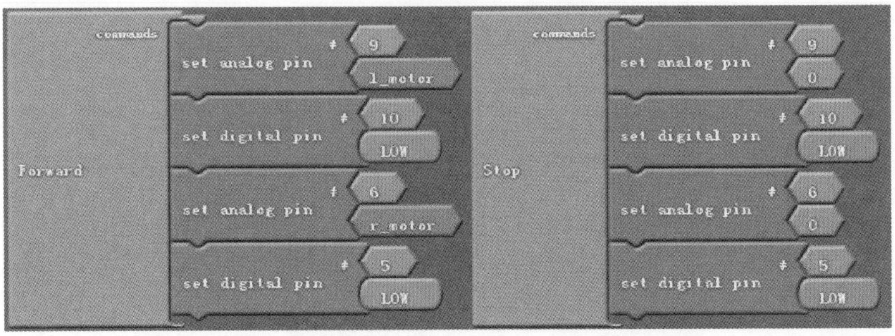

(b) Bluetooth control Micromouse to walk straight line subprogram module

Fig. 2–2–12　Bluetooth control Micromouse walk straight program (continued)

3. The flow chart of Micromouse going straight with correction (see Fig. 2–2–13)

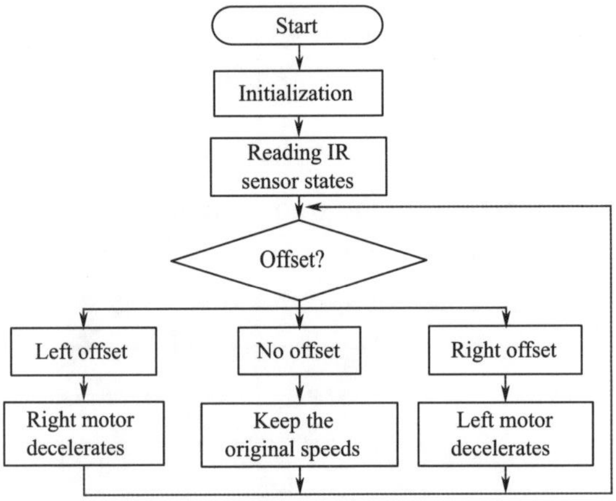

Fig. 2–2–13　Flow chart of Micromouse going straight

4. Graphical programming

Let's continue to learn how to use infrared sensors to automatically correct the posture of Micromouse.

Combined with the infrared emission intensity debugging experiment of the infrared debugging part, it can be known that:

When Micromouse is left offset, the variable L_correct is low and R_correct is high.

When Micromouse is right offset, the variable R_correct is low and L_correct is high.

Therefore, the motors' speeds can be adjusted based on these two results.

Suppose that the measured speeds of Micromouse walking straight without correction are the left motor 200 and the right motor 180. When Micromouse is shifted to the left, the right motor speed can be reduced to make Micromouse return to the center line; When Micromouse is shifted to the right, the left motor speed can be reduced to make Micromouse return to the center line. Assuming that the reduced analog quantity is 20, the programming is shown in Fig. 2–2–14.

Video

Experiment_
Micromouse
goes straight

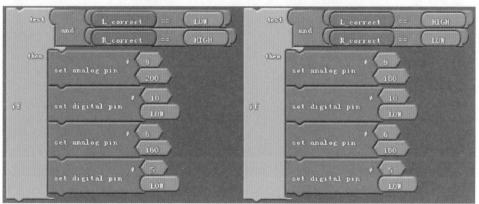

Fig. 2–2–14 Left and right correction of Micromouse

The Micromouse competition maze is 1.5 m×1.5 m in size, and the lighting conditions at each location are different. Therefore, for the integrity of the program and the occurrence of unexpected situations, the front left and right sensors are added to detect the baffle at the same time and the block cannot be detected at the same time. In the case of the board, the left and right speed of the Micromouse remains unchanged, as shown in Fig. 2–2–15.

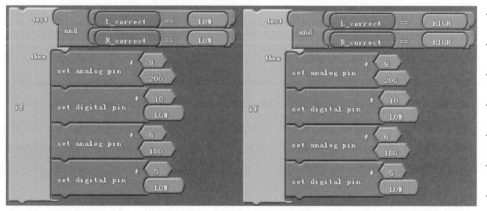

Fig. 2–2–15 Micromouse does not adjust speed in special circumstances

In order to improve the readability of the program, we set up subroutines:

Left_offset—Left offset subroutine;

Right_offset—Right offset subroutine;

Forward—forward subroutine, as shown in Fig. 2–2–16.

Now you can download the program, and then place Micromouse on the ③ and ④ positions of the test site to verify the Micromouse's self-calibration.

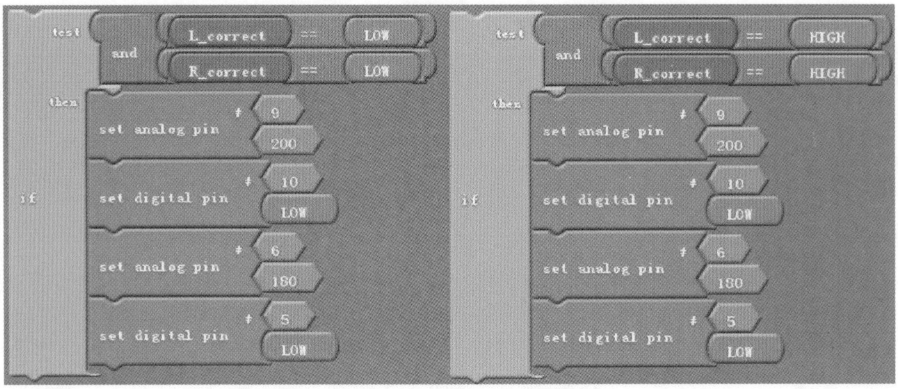

Fig. 2–2–16   Walk straight with correction

Task 2   Micromouse Turns

The turning angle of Micromouse is controlled by two parameters, namely motor speed and turning time. When Micromouse is turning, the speed of the outer motor and the rotation time of the motor (that is, the length of the delay function) work together. The larger the value, the greater the turning angle.

By matching and adjusting the speed of the outer motor and the rotation time, Micromouse can rotate any angle.

1. The flow chart of turning 90°

When Micromouse walks in a maze, the basic turning angle is 90°. The flow chart for judging whether Micromouse has turned 90° is shown in Fig. 2–2–17.

Through the ⑪, ⑫ and ⑬ marks on the test site, it can be judged whether Micromouse turns 90° accurately. Suppose it is found that Micromouse does not turn left by 90°. At this time, there are two adjustment methods (take the left turn as an example):

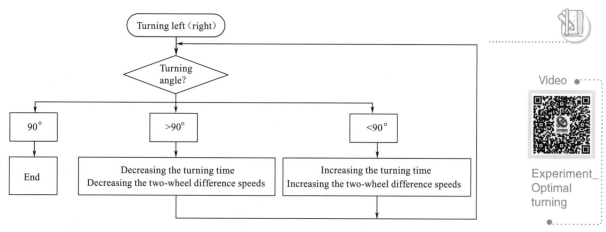

Fig. 2–2–17   The flow chart for judging the angle of Micromouse

Video

Experiment_
Optimal
turning

(1) Increasing or decreasing the analog value of pin 6.

(2) Increasing or decreasing the time of turn left.

2. Graphical programming for precise control of the turning angle

We have learned how to judge and adjust the turning angle. Next, we write the actual program.

(1) As the speed and time need to be adjusted, set the speed and time as variables and assign initial values, as shown in Fig. 2–2–18.

(2) Add Bluetooth to control the start and stop of Micromouse. If Micromouse is directly controlled by sending instructions to stop, it will introduce human error and cause some interference, so change the strategy: set a system state variable to indicate whether Micromouse is running or stopped, as shown in Fig. 2–2–19 and Fig. 2–2–20.

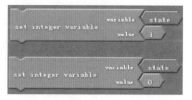

Fig. 2–2–18   Set variable program   Fig. 2–2–19   Setting status comparison
for speed and time

The next step is to control Micromouse to turn 90°. Replace the Forward subroutine with the left turn subroutine, as shown in Fig. 2–2–21.

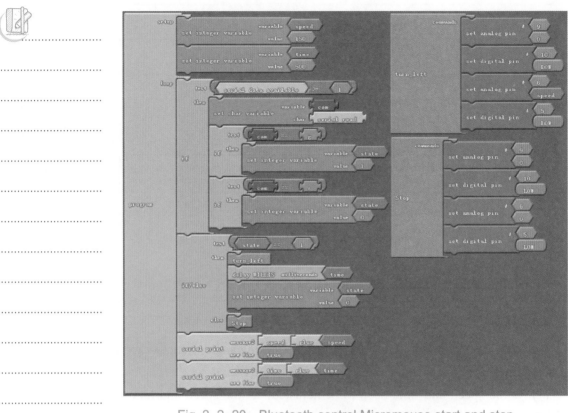

Fig. 2-2-20　Bluetooth control Micromouse start and stop

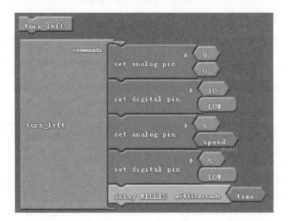

Fig. 2-2-21　Left turn subroutine

What should I do if the initial value cannot satisfy Micromouse turning 90°? The specific method is to send commands via Bluetooth to increase or decrease the size of the variable "speed" or "time", as shown in Fig. 2-2-22.

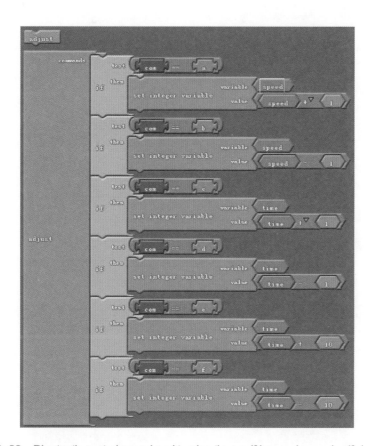

Fig. 2–2–22　Bluetooth control speed and turning time self-increasing and self-decreasing

When sending "a" "b" "c" "d" "e" "f " via Bluetooth, the speed and time will increase and decrease to obtain the accurate speed and time. Finally, don't forget to print out the speed and time through serial and record them for later use, as shown in Fig. 2–2–23.

Since there is a delay function in Micromouse turning, there is no need to add a delay function after serial printing. At this point, the speed and time required for Micromouse to turn 90° are solved.

Download the program to Micromouse and send a Bluetooth command to observe the actions of Micromouse, as shown in Fig. 2–2–24.

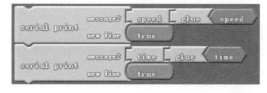

Fig. 2–2–23　Serial printing speed and time

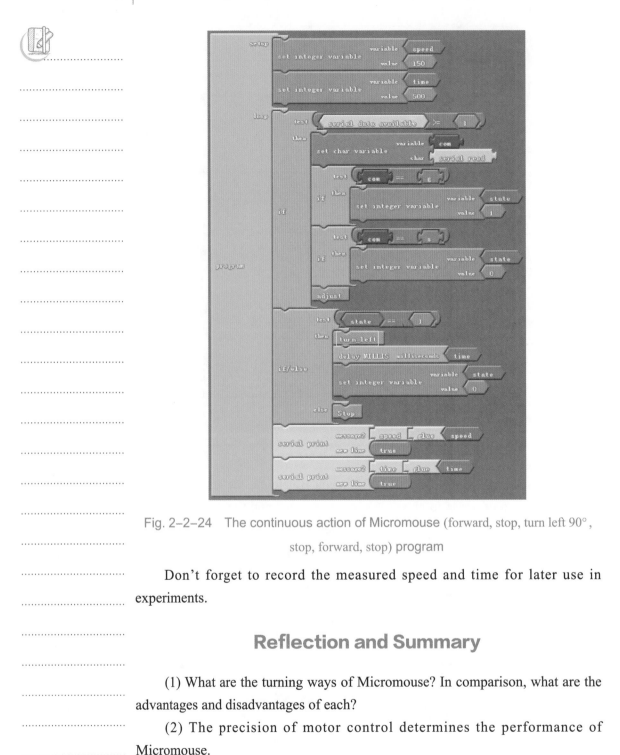

Fig. 2–2–24　The continuous action of Micromouse (forward, stop, turn left 90°, stop, forward, stop) program

Don't forget to record the measured speed and time for later use in experiments.

Reflection and Summary

(1) What are the turning ways of Micromouse? In comparison, what are the advantages and disadvantages of each?

(2) The precision of motor control determines the performance of Micromouse.

Chapter **3**

Advanced Skills and Competitions

The software and hardware of Micromouse and the basic programming and debugging methods of Micromouse have been introduced in the first two chapters. Aiming at the requirements of the China IEEE Micromouse International Invitational Competition, this chapter mainly introduces the Micromouse optimization algorithm. Master the specifications of the Micromouse competition, you can complete the maze search and optimal path selection at the fastest speed; analyze the key points of the past Micromouse maze cases, and prepare for participating in the China IEEE Micromouse International Invitational Competition.

Project 1

Analysis of Common Algorithms

 Learning objectives

Learning how Micromouse chooses paths intelligently.

How can Micromouse cross fast in a maze? The main task of Micromouse is to complete the maze search and optimal path selection according to the IEEE international standard Micromouse competition rules. It is a standard to examine the detection, analysis and decision-making capabilities of a system in an unknown environment. Let's take a brief look at this aspect knowledge.

Without predicting the path of the maze, Micromouse must first explore all the cells in the maze until it reaches the destination. Micromouse doing this process needs to know its position and posture at any time, and at the same time, it needs to record whether there are walls around the blocks it has visited. In order to save search time during the search process, Micromouse tries to avoid repeated searches that it has already been searched.

So, how to search the maze? There are usually two strategies: (1) Reach the destination as soon as possible;(2) Search the entire maze.

Both strategies have their pros and cons. Although the first strategy can shorten the searches time, it may not necessarily be able to obtain the map data of the entire maze. If the path found is not the optimal path, this will affect the time of the final sprint of Micromouse. Using the second strategy, the entire map data can be obtained, so that the optimal path can be found. However, the search time used by this strategy is longer.

Task 1   The Left-Hand Rule

In order to complete the Micromouse competition, you must know the basic method of how to search the maze. Let's start with the introduction of the left-

Video

The left-hand rule

hand rule.

In the strategy of the maze search method, Micromouse prioritizes turning left, then moving forward, and finally turning right. This strategy is called the left-hand rule. As shown in Fig. 3-1-1, the start of Micromouse in the figure. is the coordinates (0, 0), and the dashed line is the motion path of Micromouse. It can be clearly seen that whenever Micromouse encounters a branch intersection, it will select priority to turn left. Micromouse will choose to go straight when it cannot turn left, and it will turn right when Micromouse can neither turn left nor go straight.

Through the program setting, Micromouse will priority to turn left when encountering an intersection, as shown in Fig. 3-1-2.

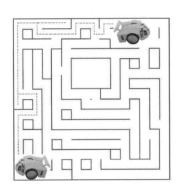

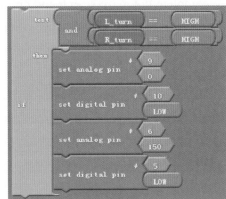

Fig. 3-1-1 The diagram of the left-hand rule

Fig. 3-1-2 Priority program for left turn

When the two rear sensors failed to detect the walls, the left wheel stops rotating, and the right wheel rotates forward to make a left turn.

Taking several key points in the schematic diagram of the left-hand rule as an example for explanation, and the corresponding decision-making is shown in Table 3-1-1. At coordinates (2, 6), Micromouse can choose to turn left or right. According to the left-hand rule, it will eventually choose to turn left. At coordinates (1, 8), Micromouse can choose to go forward or turn right, according to the left-hand rule, it will eventually choose to go forward. At coordinates (2, F), Micromouse can choose to turn right or go forward, and it will eventually choose to turn right.

Table 3-1-1   The diagram of left-hand rule key point decision-making correspondence table

Coordinates	Directions	Strategy
(2, 6)	turn left, turn right	turn left
(1, 8)	go forward, turn right	go forward
(2, F)	turn right, go forward	turn right

Task 2   The Right-Hand Rule

Video

The right-hand rule

As the name suggests, the opposite of the left-hand rule is the right-hand rule, that is, Micromouse turns right as long as there is an unpassed entrance on the right.

Similar to the left-hand rule, when Micromouse is moving forward, if there are two or more branches in the forward direction, it needs to choose which direction to turn, and the direction of the turn will cause the trajectory of Micromouse differently. The same, Micromouse prioritizes turning to the right, then going straight, and finally turning to the left. This strategy is the right-hand rule. As shown in Fig. 3-1-3, the coordinates (0, 0) in the Figure. is still the start of Micromouse, and the dotted line is still the movement path of Micromouse. The difference is that whenever Micromouse encounters a branch intersection, it will choose priority turn right. When it cannot turn right, Micromouse will choose to go straight ahead, and turn left when it can neither turn right nor go straight.

Similar with the left-hand rule, when both the left and right sensors at the rear detect intersections, according to the program, the left wheel rotates forward, the right wheel stops, and Micromouse turns right, as shown in Fig. 3-1-4.

Fig. 3-1-3   The diagram of the right-hand rule

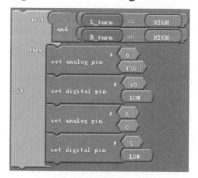

Fig. 3-1-4   Priority program for turning right

Taking several key points in the schematic diagram of the right-hand rule as an example for explanation, and the corresponding decision-making is shown in Table 3–1–2. At coordinates(0, 4), Micromouse can choose to go forward or turn right. According to the right-hand rule, it will eventually choose to turn right. At coordinates(2, 0), Micromouse can choose to turn left or right, and finally according to the right-hand rule, it will choose to turn right. At coordinates(4, 0), Micromouse can choose to turn left or go forward, and finally choose to go forward according to the right-hand rule.

Table 3–1–2　The diagram of right-hand rule key point decision-making correspondence table

Coordinates	Directions	Strategy
(0, 4)	go forward, turn right	turn right
(2, 0)	turn left, turn right	turn right
(4, 0)	turn left, go forward	go forward

Reflection and Summary

(1) What are the advantages and disadvantages of using the left-hand rule or the right-hand rule alone?

(2) The left-hand rule and the right-hand rule are set according to the priority of the turning direction. Choosing different rules when Micromouse is located at different locations will significantly improve the search efficiency.

Project 2

Advanced Control Function of Micromouse

 Learning objectives

（1）Mastering the principles and control methods of multi-sensor cooperation.

（2）Learning the principles and implementation methods of multithreading.

（3）Learning to use multi-sensor collaborative work and multi-threaded work to realize Micromouse competition.

Micromouse integrates multiple technologies and multiple disciplines, is a very complex learning platform. The means of external perception include infrared sensor detection and wireless remote control, and the movement structure is two geared motors. Since the use of wireless remote control is prohibited except for starting and stopping during the competition, whether Micromouse can reach the destination quickly and accurately depends on the detection accuracy of the sensors and the accuracy of the motors operation control. This project uses the learning of two tasks to improve the detection accuracy of the sensors and the accuracy of the motors control respectively, so that Micromouse can cross the maze quickly and accurately.

Task 1　Principle and Implementation of Multi-Sensor Cooperation

In the "Micromouse sees the world" task, we have learned how to realize sensor detection. In view of the characteristics of the infrared receiver

(IRM8601S), the closer the infrared emission frequency to 38 kHz, the farther the detection distance; on the contrary, the closer the detection distance. In this way, the detection of the walls and intersections of the maze can be realized.

The four sets of infrared sensors on the left front, right front, left rear and right rear of Micromouse respectively detect whether there are walls in the corresponding direction, and the transmission frequencies that need to be used are not the same. The infrared receiver of each group sensors uses an independent I/O interface, which can accurately receive the high and low level information in the corresponding direction (with or without walls). However, the infrared receivers use the common IO 11 interface, and it is very likely to use a certain frequency to transmit infrared light. It causes errors in the detection results of the other three directions, so how can we accurately achieve detections of walls in the four directions?

Next, we will learn to use sensors to emit alternately, multi-sensor cooperation, to realize that the four groups of sensors use their own frequency to emit infrared light and detect wall information in the corresponding direction.

1. Flow chart

Setting the variable IRNum to indicate the order of infrared emission. When the value is different, different frequencies are emitted and the corresponding infrared sensor state is called. Infrared alternate emission flow chart is shown as Fig. 3–2–1.

2. Graphical programming

(1) Program Initialization. Setting the Data_init sub-function means program digitization, where the variable IRNum ranges from 0 to 7, which is the emission sequence; F0 to F3 are the emission frequencies of the infrared sensors, as shown in Fig. 3–2–2.

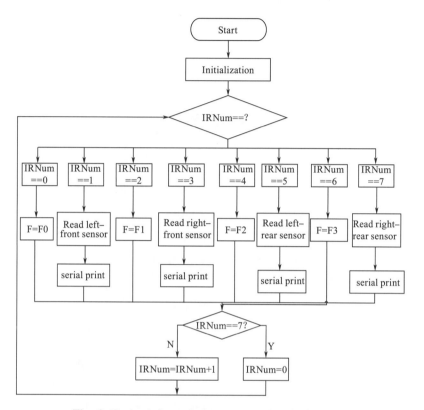

Fig. 3-2-1　Infrared alternate emission flow chart

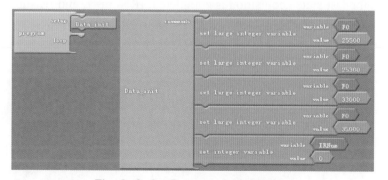

Fig. 3-2-2　Program initialization

(2) Transmitting and status acquisition of the left-front sensor. The variable IRNum is initialized to 0 when the main program "loop" runs for the first time, so the left front sensor uses F0 to emit infrared light. At the end of the program, the variable IRNum needs to be incremented by 1, so that it is convenient to run the program we need directly during the second run, as shown in Fig. 3-2-3.

When the "loop" runs for the second time, the variable IRNum=1, directly read the status of the left-front sensor, as shown in Fig. 3-2-4.

Fig. 3–2–3　The first run, emit at its own frequency

Fig. 3–2–4　The second run get the sensor status

(3) Completing the alternating cycle transmission and status acquisition of four groups of sensors. According to (2), complete the alternate emission and state acquisition of the four sensors, as shown is Fig. 3–2–5. It should be noted that when the last state acquisition is completed, IRNum=7, and IRNum needs to be re-assigned to 0 to complete the cyclic transmission and state acquisition.

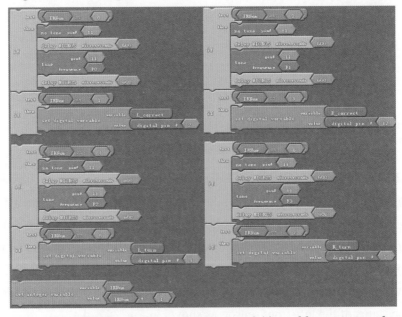

Fig. 3–2–5　Alternating emission and state acquisition of four groups of sensors

(4) Combining the program and complete the download. In order to facilitate the observation of the results, we add a "serial print" graphic after acquiring the status of each sensor, as shown in Fig. 3–2–6.

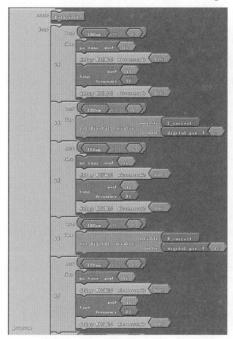 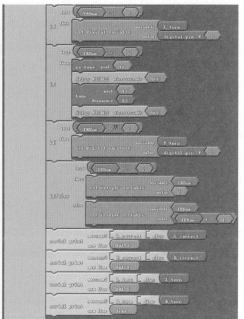

Fig. 3–2–6   The whole program of Micromouse sensors cooperation

Infrared state is shown as Fig. 3–2–7.

Fig. 3–2–7   Infrared state

Task 2　Principle and Implementation of Multithreading

We have learned the driving, speed regulation and turning of the electric motor in the basic function debugging project and attitude control project of Micromouse. Let's think about a small experiment:

The left motor rotates forward for 0.5 s, stops for 0.3 s and circulates; the right motor rotates forward for 0.4 s, stops for 0.4 s and circulates.

According to our usual thinking, can this experiment be completed? Let's have a try.

Experiment 1 Different control of left and right motors

1) Flow chart

Different control of left and right motors, as shown in Fig. 3–2–8.

```
        ┌─────────┐
        │  Start  │
        └─────────┘
             │
      ┌──────────────┐
      │ Initialization│
      └──────────────┘
             │
             ▼◄──────────┐
    ┌──────────────┐     │
    │ Left motor   │     │
    │ runs 0.5 s   │     │
    └──────────────┘     │
    ┌──────────────┐     │
    │ Left motor   │     │
    │ stops 0.3 s  │     │
    └──────────────┘     │
    ┌──────────────┐     │
    │ Right motor  │     │
    │ runs 0.4 s   │     │
    └──────────────┘     │
    ┌──────────────┐     │
    │ Right motor  │     │
    │ stops 0.4 s  │─────┘
    └──────────────┘
```

Fig. 3–2–8　Different control of left and right motors

2) Graphical Programming

(1) Program initialization. The forward rotation and stop of the left and right motors are shown in the Fig. 3–2–9.

Fig. 3–2–9　Program initialization

(2) Add delay. Adding delays separately and name the subroutines as "left_motor" and "right_motor", as shown in Fig. 3–2–10.

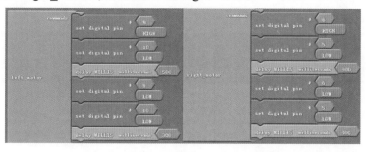

Fig. 3–2–10　Name the subroutines

(3) Combining and download the program(see Fig. 3–2–11). Now observe the rotations of Micromouse. The result doesn't meet the experimental requirements.

Fig. 3–2–11　Combining program

Delay: Keep the current output state of the controller unchanged, and continue to execute the following program after a certain period of time.

The program is executed sequentially from the top to the bottom. The "delay" function will take effect on the entire system, which will cause the subsequent code execution to lag. The main program is too long will cause the real-time performance of the system to deteriorate, thus making the control of the motors difficult. Below we learn a multi-threaded control method to improve the control accuracy of the motor.

Experiment 2 Multithreading control motor operation

Multithreading refers to a technology that implements concurrent execution of multiple threads from software or hardware. It can also be simply understood as the simultaneous execution of multiple main programs, except for variable calls, and no interference between multiple threads

1) Flow chart

The flow chart of multithreading is shown in Fig. 3–2–12.

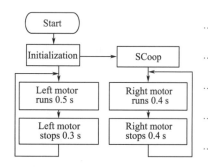

Fig. 3–2–12　The flow chart of multithreading

2) Graphical programming

The realization of multithreading function needs to use the graphics in the "SCoop" column, as shown in Fig. 3–2–13.

Fig. 3–2–13　Multithreading graphic

The "Sleep" delay in "SCoop" only affects the multithreading itself, not the main function. "Delay" has no effect on the multithreading.

(1) Initialize multithreading and add delay. Move the right motor to the multi-thread and replace the delay with the "Sleep" function, as shown in Fig. 3–2–14.

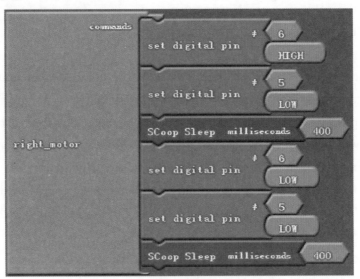

Fig. 3–2–14　Adding SCoop Sleep

(2) Combining and download programs, as shown in Fig. 3–2–15.

Fig. 3–2–15　Multithreading operation

It can be found by running the program that the movements of the left and right motors have reached expectation without interference.

Task 3   Micromouse Avoids Obstacles and Runs Flexibly

In this task, we will use the methods that we have learned to improve the accuracy of sensor detection and the accuracy of motor operation control to implement Micromouse to walk the maze quickly and accurately.

The infrared sensor emission and status acquisition are placed in the main program, and the motor control (speed regulation and turning) is placed in multiple threads. The main program and multithreading run in parallel, without interfering with each other except variable calls; the real-time performance of the program is improved.

1) Flow chart

The left-front and right-front sensors are used to correct the posture during operation, so:

When the left front sensor detects the wall, the right motor decelerates, otherwise it maintains high speed.

When the right front sensor detects the wall, the left motor decelerates, otherwise it maintains high speed.

The left-rear and right-rear sensors are used to detect intersections during operation, so:

When the left rear sensor detects the intersection, turn left, otherwise go straight.

When the right rear sensor detects the intersection, turn right, otherwise go straight.

Note: When entering any turn, you need to switch the multi-threading state from the straight-line correction state to the turning state, and reset the multi-thread state back to the straight-line correction state after the turn.

The flow chart is shown in Fig. 3–2–16 and Fig. 3–2–17.

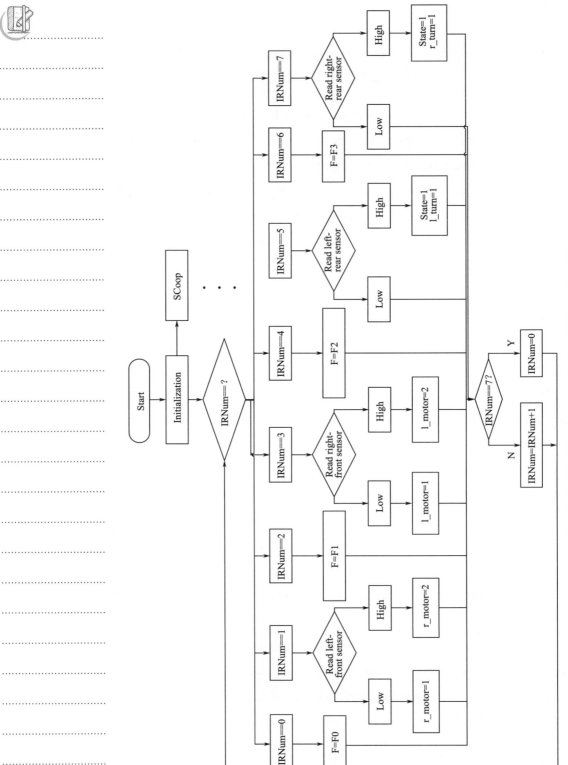

Fig. 3-2-16 The flow chart of multithreading(1)

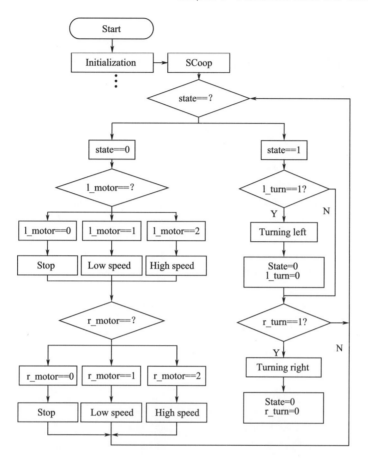

Fig. 3–2–17   The flow chart of multithreading(2)

2) Graphical programming

(1) Initializing all system variables, as shown in Fig. 3–2–18.

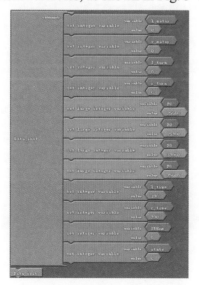

Fig. 3–2–18   Initializing

The variables in the Fig. 3–2–18 will be called or assigned when the program is running. The value ranges and meanings are shown in Table 3–2–1.

Table 3–2–1　The value ranges and meanings

Item	Meaning	Ranges		
		0	1	2
r_motor	Right motor control	Stop	Low speed	High speed
l_motor	Left motor control	Stop	Low speed	High speed
state	Multithreaded state	Correction to go straight	Turn	No
l_turn	Left turn control	Straight	Turn left	No
r_turn	Right turn control	Straight	Turn right	No
F0~F3	Four groups of transmit frequency	Micromouse eyes to see the world task to obtain		
l_time	Time required to turn left 90°	Micromouse learns to turn task to obtain		
r_time	Time required to turn right 90°	Micromouse learns to turn task to obtain		

Because of the parameters of each Micromouse are different, users must replace them with their own measured data when debugging.

(2) Calling the multi-sensor cooperation program and add variable assignment, as shown in Fig. 3–2–19 to Fig. 3–2–22.

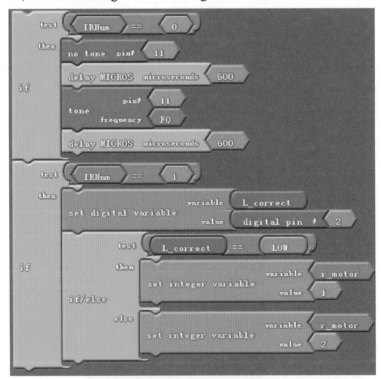

Fig. 3–2–19　The left-front sensor

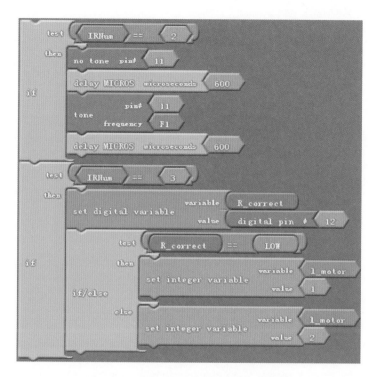

Fig. 3-2-20   The right-front sensor

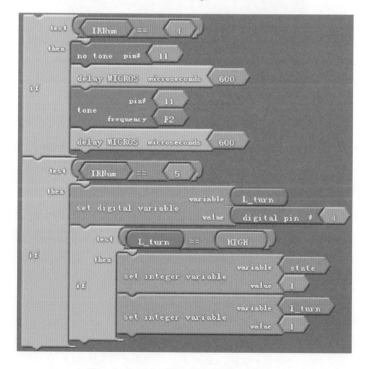

Fig. 3-2-21   The left-rear sensor

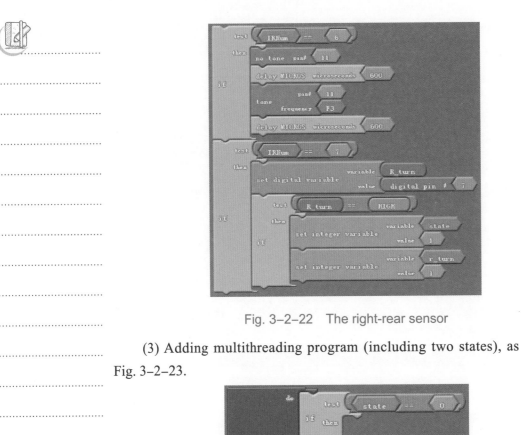

Fig. 3-2-22   The right-rear sensor

(3) Adding multithreading program (including two states), as shown in Fig. 3-2-23.

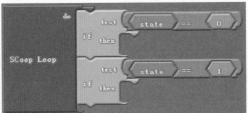

Fig. 3-2-23   Multithreading with two states

The two "if " graphics can also be replaced with one "if/else", but in order to optimize the function of the program in the future, it is recommended to keep the two "if " graphics.

(4) Motors control in the correct straight line state, as shown in Fig. 3-2-24.

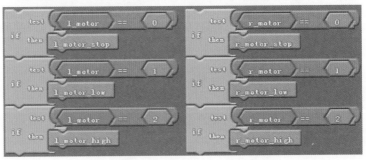

Fig. 3-2-24   Motors control

In Fig. 3–2–25, the high-speed values (150, 165) of the left and right motors are the data measured in the walking straight without correction task, and the low-speed values (120, 143) are the data measured in the Micromouse runs task.

Fig. 3–2–25　Speeds control（low and high）

Stop program is shown as Fig. 3–2–26.

Fig. 3–2–26　Stop

(5) Motors control in the turning state. In order to show the turning process clearly, we add 0.2 s stop procedures before and after the turn (see Fig. 3–2–27).

Because of Micromouse turns in pirouetting, when the turning is over, its rear sensors are still in the intersection, so add a 0.35 s forced straight program to make Micromouse walk through the current intersection and avoid detection errors (see Fig. 3–2–28).

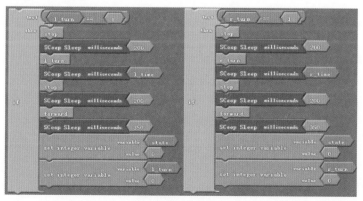

Fig. 3–2–27　Motors control in turning state

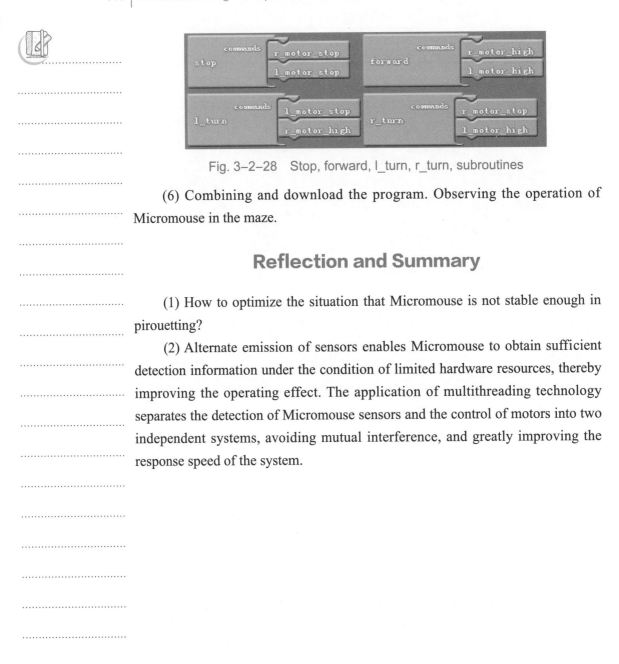

Fig. 3-2-28   Stop, forward, l_turn, r_turn, subroutines

(6) Combining and download the program. Observing the operation of Micromouse in the maze.

Reflection and Summary

(1) How to optimize the situation that Micromouse is not stable enough in pirouetting?

(2) Alternate emission of sensors enables Micromouse to obtain sufficient detection information under the condition of limited hardware resources, thereby improving the operating effect. The application of multithreading technology separates the detection of Micromouse sensors and the control of motors into two independent systems, avoiding mutual interference, and greatly improving the response speed of the system.

Chapter **4**

Extended Application

This chapter introduces the use of Micromouse core technology to cultivate students' ability of independent inquiry and hands-on solution. Master how to combine and expand functions flexibly on the basis of basic technology. To cultivate students' design and practice ability and train their spirit of "innovation, creativity and creation".

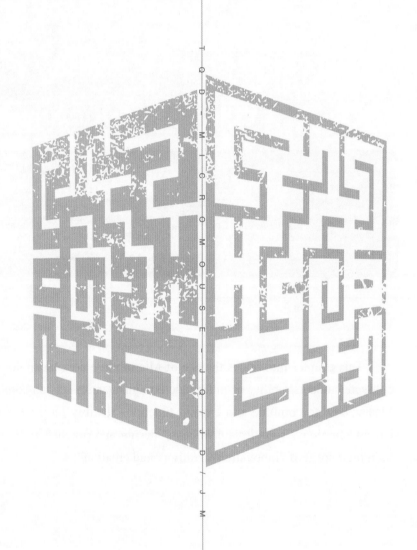

Project 1

Structural Composition of TQD-IOT Engineering Innovation Course Platform

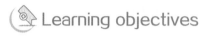 Learning objectives

(1) Understanding the relationship between engineering innovation course platform and Micromouse.

(2) Learning the hardware composition of TQD-IOT engineering innovation course platform.

(3) Learning to use TQD-IOT engineering innovation course platform for typical IOT application development.

Micromouse involves a variety of technologies, which can be easily extended to other scenes. This project mainly introduces the TQD-IOT engineering innovation course platform based on the technology of Micromouse.

Task 1　Relationship Between Engineering Innovation Course Platform and Micromouse

Through the above basic knowledge learning, readers have mastered the use of Arduino software and hardware, Bluetooth, infrared sensor and motor. TQD-IOT engineering innovation course platform, as shown in Fig. 4-1-1, uses the same controller as Micromouse, and makes a lot of supplements in actuator and sensor. Table 4-1-1 shows the comparison between the two.

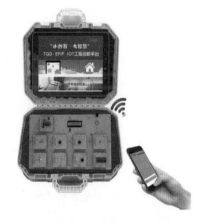

Fig. 4-1-1　TQD-IOT engineering innovation course platform

Table 4-1-1　The comparison between TQD-IOT engineering innovation course platform and Micromouse

Comparison items	TQD-IOT engineering innovation course platform	Micromouse competition platform	Comparison results
Development	Arduino	Arduino	Same
MCU	ATmega328P	ATmega328P	Same
Actuator	DC motor, steering gear	DC motor, stepping motor, deceleration motor, servo motor, etc.	Increment
Sensor	There are more than 30 kinds of sensors，such as infrared sensor, ultrasonic sensor, vibration sensor, photosensitive sensor, LED, LCD, dot matrix module, Bluetooth module, etc	IR sensor	Decrement
Intelligent algorithm	Non	The right-hand rule, the left-hand rule, the central rule, etc.	Different

Through comparison, it can be seen that TQD-IOT engineering innovation course platform and Micromouse have the same points and different points. It is the continuation and expansion of Micromouse knowledge, and has complementary and progressive complementary roles for learning electronic system design and application.

The features of TQD-IOT engineering innovation course platform are as follows:

1) The system uses international open source Arduino software and hardware platform

The CPU uses a high-performance, low-power AVR ATmega328P microcontroller with a built-in Bluetooth module and supports up to 6 analog and 14 digital output (compatible with PWM output).

2) International interesting graphical programming, easy to learn and understand

The platform software supports both code and graphical programming, in which graphical programming can be converted into code in real time, thus helping students learn C language and other programming modes. Graphical programming, the variety of modules, the combination of graphics using color modules, to improve the fun of programming.

3) Multimedia App online learning resources are rich and diverse

Multimedia App for platform, which is its own intellectual property software, including experimental software and online learning resources. The experimental software is set for each chapter as experimental purpose, software simulation, experimental equipment, hardware connection, task programming, effect demonstration, after-class thinking. Online learning resources enable students to learn and know their use, use and know their place, know their generation, know their origin and derive from it through modern intelligent distance learning. Covers all teaching resources (detailed explanations, procedures, pictures, videos in each chapter), allowing students to learn freely whenever and wherever they want. Use three-dimensional teaching methods to maximize the motivation of students.

4) IOT modular design, powerful extension

The platform is modular and can be divided into three areas: control, display and operation. Optional IOT modules include: AVR control module, Bluetooth communication module, monochrome and multi-color LED module, key module, dot matrix module, ultrasonic module, active and passive buzzer module, temperature sensor module, photosensitive sensor module, fan motor module, infrared counter-fire module, speech recognition module, etc., covering typical Internet of things core sensor and executor applications. It can easily implements analog intelligent control of various smart home.

5) Design of modular magnetic absorption structure is safe, reliable and convenient

Each module of the platform is designed with magnetic absorption structure. The multi-contact magnet spring pins have close contact, no wires need to be connected, strong adsorption, strong contact, safe and convenient. The electrical connection is designed with 20 pin spring pin and groove touch. The connection diagram of innovative training platform is shown in Fig. 4–1–2.

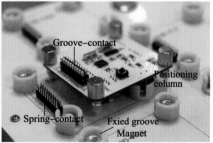

Fig. 4–1–2　Connection diagram

The platform 2×4 operation area supports 8 IOT universal module linkage experiments at the same time. The structure is convenient for students to plug, improves the reliability of the connection and prolongs the life of the product.

6) Wireless Bluetooth communication, convenient, fast and effective

The control module integrates a wireless Bluetooth module, with a communication distance up to 10 m, and a separate Bluetooth APP for mobile phones, which facilitates data transmission between users and devices.

Task 2　Hardware Composition of Engineering Innovation Course Platform

The engineering innovation course platform is divided into three parts, control area, display area and execution area. Modules can be placed in each area. The module and the platform use the magnetic absorption design, which is patented by the state utility model, that is, put-through, no wire connection is required, the experimental process is more convenient, up to nine operational positions, and various DIY experimental designs can be easily achieved.

1. Control area

Used to place the controller module. The core controller adopts the same technology as Micromouse, as shown in Fig. 4–1–3. The control consists of a digital I/O interface and an analog I/O interface. Through its upper 20 pin port (also known as antenna thimble), it can be easily connected to different sensors, actuators and drivers.

 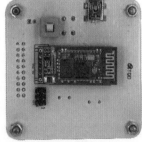

Fig. 4–1–3　Core controller

The components are integrated in the controller module, which is convenient for management and operation. Users can connect the experimental platform with the user's mobile phone through Bluetooth. During the experiment, the user can directly use the mobile phone to control the experimental platform at the remote end, saving cost and occupying space.

(1) Bluetooth module is integrated in the core controller. Users can operate the experimental steps through the Bluetooth serial port software installed on the mobile phone, and also receive the status information fed back by the module, so as to realize the wireless transmission of information.

(2) The back of the core controller is equipped with AVR main control chip, crystal oscillator, multiple groups of capacitor components and resistance components. All the core components are integrated in the back of the circuit board, so secondary connection is not required. At the same time, the reset key is installed to reset and clear the program, which is convenient to continue the new experiment.

(3) The front of the core controller is equipped with a one button switch, which can control the working state of the module by pressing one key without any other operation. The USB interface close to the edge can be directly connected to the computer, without the need to connect other downloaders. The program is written into the main control chip to control the experimental box.

Arduino hardware platform is based on AVR, and the AVR library is compiled and packaged for the second time. Registers and address pointers have been set, so there is no need to make special settings when using.

Arduino hardware platform has the following main features:

Processor: Atmel Atmega 328P.

Digital I/O: Digital input/output port D0-D13.

Analog I/O: Analog input/output port A0-A5.

Support ICSP download, support TX/Rx.

Input voltage: USB interface power supply or 5-12 V external power supply.

Output voltage: Support DC 3.3 V and 5V output.

Among them, D3, D5, D6, D9, D10, D11 of digital interface (see Fig. 4–1–4) can also be used as PWM output interface.

In addition to D0 and D1 used for

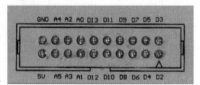

Fig. 4–1–4　Interface

serial communication, all other pins of the control module are led out, which is convenient for free combination of experiments.

In daily life, there are many devices, such as notebook computers, speakers, headphones, etc., are built-in Bluetooth module, in order to achieve the function of wireless transmission. The Bluetooth module model used in the experiment is HC-06, as shown in Fig. 1–2–5. The communication distance can be as far as 10 m, which greatly facilitates the wireless transmission of data between users and equipment.

2. Display area

Many electronic products, such as calculators, electronic watches, multimeters and so on, contain LCD module, which has the characteristics of high display quality, small size, light weight and low power consumption. This platform adopts 1602 LCD module, as shown in Fig. 4–1–5. It can display up to 2×16 characters, with adjustable definition, low delay, good real-time performance, and can accurately display all kinds of data or information.

Fig. 4–1–5   LCD module

Pin connection between LCD module and core controller is shown in Table 4–1–2.

Table 4–1–2   Pin connection between LCD module and core controller

No.	LCD module	Core controller
1	VCC	VCC
2	GND	GND
3	RS	D12
4	EN	D11
5	IO	D5
6	IO	D4
7	IO	D3
8	IO	D2

These pins are fixed and cannot be changed, when using the LCD module, these pins can no longer be used.

3. Execution area

The platform provides 2×4 operable module areas, as shown in

Table 4–1–3. According to the actual needs of the experiment, the type or quantity of modules can be increased or decreased freely to realize "innovation, creativity and creation". All modules are connected by magnetic adsorption, which can effectively avoid the damage of the equipment.

Table 4–1–3　Operable modules

Single-color LED	Tri-color LED	Key module	Photosensitive sensor
Lattice display module	Ultrasonic sensor	Temperature sensor	Fan module
Passive buzzer	Active buzzer	Infrared receiver	Infrared transmitter

Reflection and Summary

(1) What are the similarities and differences between TQD-IOT engineering innovation course platform and Micromouse platform?

(2) What are the hardware components of TQD-IOT engineering innovation course platform? What software is used for programming?

(3) Each module of TQD-IOT engineering innovation course platform adopts magnetic adsorption structure design, and is equipped with independent mobile phone Bluetooth APP to facilitate data transmission between users and devices.

Project 2

IOT Extended Application of Micromouse Technology

Learning objectives

(1) Learning the intelligent control of lighting and the common control method of multiple signals.

(2) Learning the principle and implementation of common security.

(3) Understanding what visual persistence is.

(4) Try DIY image display.

This project uses TQD-IOT engineering innovation course platform to simulate the application of Micromouse technology in real life through the three tasks of "Micromouse controller controls the lighting system", "Micromouse controller controls the security system" and "Micromouse controller controls the display system".

Task 1 Micromouse Controller Controls the Lighting System

In recent years, with the development and maturity of lighting construction, intelligent control system is gradually introduced into lighting management system. The intelligent lighting system is a system for intelligent control and management of lighting. Compared with traditional lighting, it can realize the management of soft start, dimming, one key scene, one-to-one remote control and all on/off of partition lighting. It can be controlled by remote control, timing, centralized and remote control, and even advanced intelligent control of lighting by computer to realize energy saving, environmental protection, comfort and convenience.

Subtask 1　Color Transformation of Light

1. Application scenarios

Lights can be seen everywhere, such as fluorescent lamps for classroom lighting, red warning lights, yellow fog lights on cars, etc. There are also many lights with changing colors, such as red, yellow and green lights to indicate traffic, as shown in Fig. 4–2–1, neon lights used for decoration in shopping malls. Lighting is the beautician of the city. When the night falls, the colorful lights make the city more beautiful. So, how are these different color lights realized? How do different colors switch?

Fig. 4–2–1　Three color traffic lights

2. Hardware introduction

1) LED introduction

LED is called light emitting diode in English. It is a solid-state semiconductor device that can convert electrical energy into visible light. In daily life, LED is widely used. When doing the experiment, the commonly used LED shape is shown in Fig. 4–2–2.

LED has a variety of dimensions, it has a single guide electricity, plus the appropriate forward voltage, can light. According to the different materials or filling gas, it can emit red, green, blue, yellow and other colors of light. Different colors of LED mixed with each other can form other colors, such as two-color lights, three-color lights and even full-color lights.

The mixing principle of three primary colors is shown in Fig. 4–2–3.

Fig. 4–2–2　Commonly used LED shape

Fig. 4–2–3　Three primary color mixing principle

2) Module introduction

Tri-color LED module: The tri-color LED used in the experiment is common anode design, as shown in Fig. 4–2–4.

Fig. 4–2–4　The tri-color LED

The tri-color LED module can be understood as red, green, blue three single color LED lights placed in a cover. When the red LED is on, it will emit red light; when the green LED is on, it will emit green light; when the blue LED is on, it will emit blue light; when it is mixed according to different types, it can emit light of other colors. Three groups of monochromatic lamp circuit can be controlled separately, when the input high level is on, the input low level is off.

As shown in Table 4–2–1, the conversion of different colors can be realized by controlling the high and low levels of the three pins of the tri-color LED module.

Table 4–2–1　The tri-color LED module pins

No.	The tri-color LED pins	Core controller
1	VCC	VCC
2	R	D9 (Red primary color)
3	G	D10 (Green primary color)
4	B	D13 (Blue primary color)

3. Operation steps

Next, through a tri-color LED module and key module to introduce how to create a colorful atmosphere.

1) Flow chart

Before programming, think about the overall process: The tri-color LED module can light up different colors and change colors in the order of red, green and blue, as shown in Fig. 4–2–5.

2) Graphic programming

(1) First, select the main program module in the "control" column, as shown in Fig. 4–2–6.

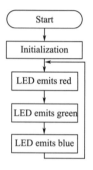

Fig. 4–2–5   Flow chart of creating
color atmosphere

Fig. 4–2–6   The main program graphic

(2) Set the pin level of tri-color LED module, as shown in Fig. 4–2–7.

Fig. 4–2–7   Two kinds of digital pin graphics

Note: The first one is to read the status, so the sensor uses this figure; the second is to set the high and low level of the pin, so the actuator selects this figure.

As shown in Fig. 4–2–8, when IO 9 outputs high level, IO 10 and IO 13 output low level, it will emit red.

Fig. 4–2–8   Red, green and blue

When IO 10 output high level, IO 9, IO 13 output low level, emit green.

When IO 13 output high level, IO 9, IO 10 output low level, emit blue.

(3) Color change. Think about it, can you directly combine the above programs to realize the transformation of three colors?

As shown in Fig. 4–2–9, after downloading the program, we found that the tri-color LED lights up white. This is because the program runs very fast, for the human eye is almost at the same time received red, green and blue, the proportion of three colors mixed to become white.

Therefore, it is necessary to use some means to distinguish the colors, so that the three colors emit longer time.

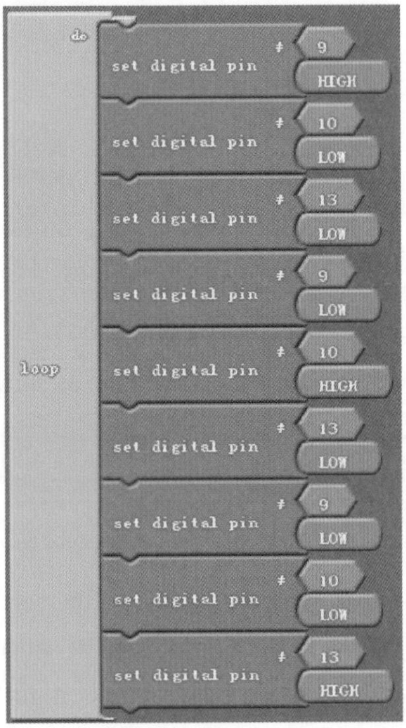

Fig. 4–2–9   Direct combination of three colors

The delay graph in the "control" column can meet this purpose, as shown in Fig. 4–2–10.

As shown in Fig.4–2–11, the color conversion of tri-color LED module controlled by key is completed.

Fig. 4–2–10   delay

Now upload the program, observe the effect. In addition, readers can change the high and low levels of pin 9, pin 10, pin 13 to observe the color change of three color LED lights.

Note: When uploading the program, the white button on the control module must be pressed down.

Expanded training: How to realize the color beyond red, green and blue? (Tip: The color can be mixed of two above.)

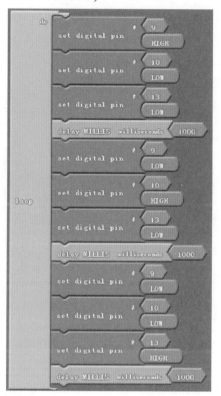

Fig. 4–2–11　Tri-color conversion program

Subtask 2　Mobile phone remote-control lights

1. Application scenarios

With the acceleration of the pace of daily life, in daily life, instant operation has been unable to meet our pursuit of life. A method that can be remotely operated anytime, anywhere has become a desired goal.

With the development of Internet of things, cloud computing, wireless communication and other new technologies, smart home has been developed rapidly, which makes people's life more convenient, as shown in Fig. 4-2-12. Users use smart phones to control the devices in their home to realize remote control, scene control, linkage control and timing control. When we go to work, we can remotely monitor the situation at home and find out the situation of the elderly and children in time; after work, we only need to send remote instructions to control the air conditioner to start, the rice cooker to cook, and the water heater to turn on. In this way, when we get home, we don't need to spend any more waiting time, and we can enjoy a full life.

Fig. 4-2-12　Smart home controlled by mobile phone

2. Hardware introduction

1) Wireless communication principle

Wireless communication is a kind of communication mode which uses the characteristics of electromagnetic wave signal in free space to exchange information. Wireless communication technology mainly includes radio communication, microwave communication, infrared communication and optical communication. Among them, radio communication is the most widely used. It is a kind of communication mode that uses electromagnetic wave signal to transmit information in free space.

Bluetooth technology is a common wireless communication (data) transmission mode. It has the characteristics of low cost, low power and strong openness. It can be seen in cars, mobile phones and computers.

2) Module introduction

Bluetooth module has been integrated into the control module to avoid wiring errors. Bluetooth module has been described in detail in the previous article, and will not be repeated here.

The single-color LED module is very similar to the tri-color LED module. The single-color LED can only emit one color, as shown in Fig. 4–2–13. Different colors can be emitted according to the difference of filling gas. The pins connection between the single-color LED module and the core controller is shown in Table 4–2–2.

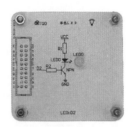

Fig. 4–2–13 Single-color LED module

Table 4–2–2 Pins connection between single-color LED and core controller

No.	single-color LED pins	Core controller
1	VCC	VCC
2	GND	GND
3	IO	D2

3. Operation steps

Next, through the mobile phone Bluetooth App to control the LED on and off, so as to learn wireless communication control.

1) Flow chart

Before programming, consider the overall process: first, the mobile phone debugging tool (mobile phone Bluetooth App) should be connected (paired) with the Bluetooth integrated in the control module; then the App sends instructions, and the control module reads these instructions to control the LED module on and off, as shown in Fig. 4–2–14.

2) Graphic programming

(1) The connection between Bluetooth APP of mobile phone and Bluetooth module is shown in Fig. 4–2–15.

Search for a new device and pair it with the control module Bluetooth (for pairing key, please check the sticker or try the default key "1234").

(2) Bluetooth control LED on and off. The experiment of Micromouse wireless control has been introduced before. The experiment is carried out according to the knowledge learned, as shown in Fig. 4–2–16.

In order to make the experimental process more clear and intuitive, modify the instructions sent by the mobile phone Bluetooth APP and then returned information:

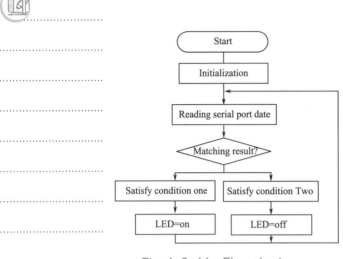

Fig. 4–2–14 Flow chart

Fig. 4–2–15 The connection between app and Bluetooth module

Send the command "o", corresponding to open, that is to turn on the LED, and the returned information is modified to "LED is open!"

Send the command "c", corresponding to close, that is to turn off the LED, and the returned information is modified to "LED is close!"

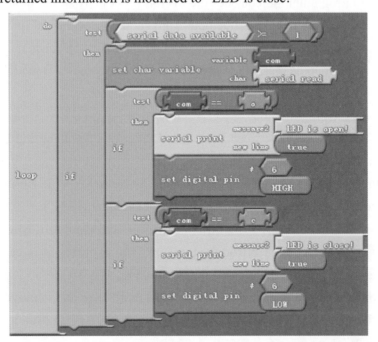

Fig. 4–2–16 Bluetooth communication experiment

Note: Because USB serial port and Bluetooth communication share the same RXD and TXD, the white button on the control module must be pressed down when uploading the program. When Bluetooth communication is completed after uploading, the button must be popped up.

Expanded training: The above experiment has realized the remote control of the light on and off. Can we add new functions to realize the remote control and automatic adjustment of the light to realize the light on and off and the light and dark? (Tips: Use photosensitive sensor.)

Task 2   Micromouse Controller Controls the Security System

With the economic development and social progress, people's living standards have been greatly improved. After enjoying life, home safety has become a matter of great concern to people. Security system is an important technical means to implement security control. In the current situation of security demand expansion, its application in the field of security technology prevention is more and more extensive.

With the development of photoelectric information technology, microelectronics technology, micro computer technology and video image processing technology, the traditional security system is also gradually moving from digital and network to intelligent. This kind of intellectualization means that the system can automatically detect and identify the abnormal situation in the monitoring screen without human intervention, and can make early warning/alarm in time when there is any abnormality.

Intelligent security system can be simply understood as: image transmission and storage, data storage and processing, accurate and selective operation of the technical system. A complete intelligent security system mainly includes access control, alarm and monitoring. The biggest difference between intelligent security and traditional security is intelligence. China's security industry is developing rapidly and more popular, but the traditional security is more dependent on people, which is very labor-intensive, and intelligent security can realize intelligent judgment through machines, so as to achieve what people want to do as much as possible.

In recent years, campus security has been paid much attention by education and law enforcement departments at all levels. To speed up the construction of the campus security system and improve the ability to prevent and stop crimes and respond to emergencies are effective means to avoid campus security incidents. Campus security system has the function of preventing accidents. Through various technical management means, the campus security risks can be predicted in advance to effectively avoid the occurrence of campus safety accidents. Based on the analysis of the actual situation of the campus, the campus security system uses relevant scientific and technological means to make the security system cover the whole campus and surrounding areas, and effectively help the school to do a good job in campus security management, as shown in Fig. 4-2-17.

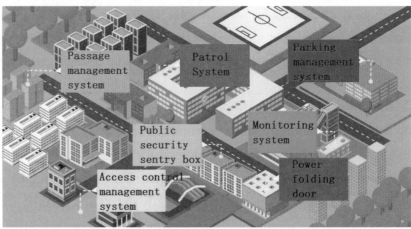

Fig. 4-2-17　IOT security system

Subtask 1　Application of Electronic Fence

1. Application scenarios

In real life, many cases will use ultrasound as a judgment condition to judge whether there is an object near.

Scene 1: Application of ultrasonic sensor in motion detection.

Ultrasonic sensors have been widely used in automatic detection of moving objects in recent years. Ultrasonic sensor mainly uses Doppler effect to work. It emits high-frequency ultrasound which is beyond the human body's perception through crystal oscillator. Generally, 25-40 kHz wave is selected. Then the control module detects the frequency of reflected wave. If there is object movement in the area, the reflected wave frequency will fluctuate slightly, that is Doppler effect, so as to judge the object movement in the lighting area in order to control the switch.

Scene 2: Ultrasonic sensor is used for vehicle detection of brake system.

Sensors are installed in parking lots and garages to detect when you are close to walls or other obstacles in the garage, as shown in Fig. 4–2–18. If the entrance is controlled by the gate system, when there is a vehicle under the railing, the railing cannot be lowered. Ultrasonic sensors are particularly suitable for controlling this process. They detect objects that are not affected by the type or color of the vehicle and monitor the entire area under the railing.

Fig. 4–2–18   Ultrasonic testing

2. Hardware introduction

1) Introduction of sound wave

Sound wave is the transmission form of sound, which belongs to mechanical wave. The propagation of object vibration in air or other media is called sound wave. The transmission of sound wave must depend on some medium. It is impossible to transmit sound wave in vacuum. Different media have different transmission speed. Sound wave has not only speed, it also has frequency, which is the number of times an object vibrates in unit time. Some of the sounds we heard were pleasant, some deep, and some sharp.

The frequency of sounds varies greatly, which makes sound waves colorful.

These are all sounds that we can hear. There are also many "inaudible sounds" in nature, as shown in Fig. 4–2–19.

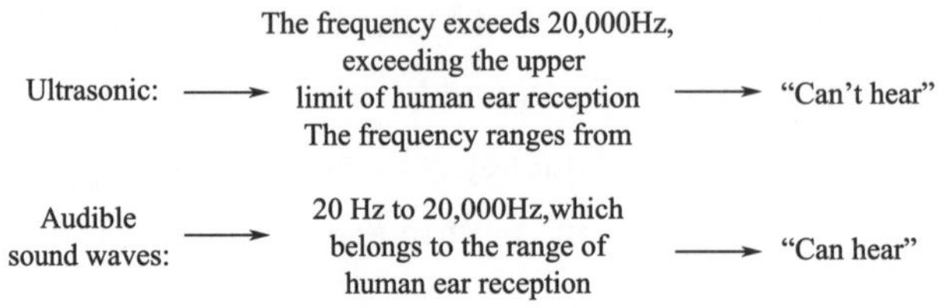

Ultrasonic: ⟶ The frequency exceeds 20,000Hz, exceeding the upper limit of human ear reception The frequency ranges from ⟶ "Can't hear"

Audible sound waves: ⟶ 20 Hz to 20,000Hz,which belongs to the range of human ear reception ⟶ "Can hear"

Fig. 4–2–19   Difference between ultrasonic wave and audible sound wave

According to the different frequencies of sound waves, they can be divided into the following categories:

The sound wave with frequency lower than 20 Hz is called infrasonic wave or ultra-low sound wave.

The sound wave with frequency from 20 Hz to 20 kHz is called audible sound wave.

The sound wave with frequency from 20 kHz to 1 GHz is called ultrasonic wave.

The sound wave whose frequency is more than 1 GHz is called special ultrasonic or microwave ultrasonic.

2) Module introduction

The ultrasonic sensor adopts the integrated module of transmitter and receiver, as shown in Fig. 4–2–20.

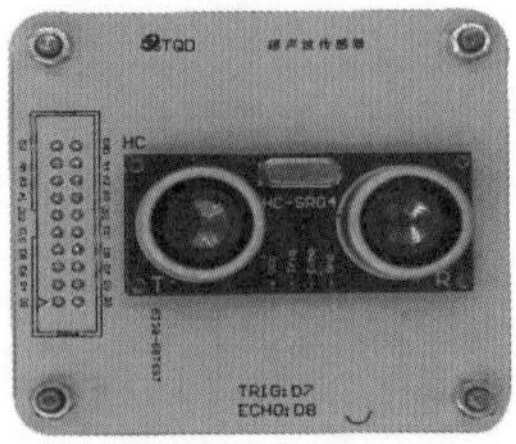

VCC and GND are the power supply and grounding terminals in the pins of ultrasonic sensor.

Fig. 4–2–20   Integrated transmitter and receiver module

Trig is the trigger signal input (ultrasonic emission when there is signal input).

Echo is the echo output (when the ultrasonic sensor receives the echo, output signals). The pins connection between ultrasonic module and core controller is shown in Table 4–2–3.

The output value of ultrasonic sensor is in cm.

The differences between digital and analog pins have been described in the previous tasks. Please guess, the ultrasonic sensor need to use analog pins or digital pins?

Table 4-2-3   Pins connection between ultrasonic module and core controller

No.	Ultrasonic module pins	Core controller
1	VCC	VCC
2	GND	GND
3	Trig	D7
4	Echo	D8

The distance measured by ultrasonic wave is constantly changing, but it only records whether the ultrasonic wave is emitted and received, and then calculates the distance according to the formula (time difference multiplied by the sound speed). Therefore, the ultrasonic sensor uses digital pins.

In addition, the tri-color LED module will be used in the experiment, which has been explained in the color conversion experiment and will not be repeated here.

3. Operation steps

Next, through the electronic fence experiment to introduce the application of ultrasonic distance measurement in intelligent security.

1) Flow chart

Before programming, first think about the overall process: Ultrasonic module as a detection device, first detects the distance of obstacles, and then judges the distance. If it is greater than 15 cm, it is a safe distance; if it is less than 15 cm, an alarm will be given, and the tri-color LED will be bright red, as shown in Fig. 4-2-21.

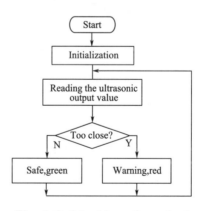

Fig. 4-2-21   Alarm flow chart

2) Graphic programming

Experiment 1: Ultrasonic distance measurement.

(1) First, find the 1602 LCD module, as shown in Fig. 4-2-22.

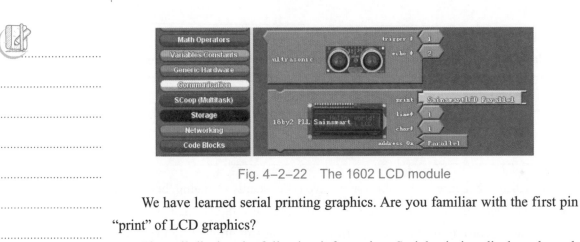

Fig. 4-2-22　The 1602 LCD module

We have learned serial printing graphics. Are you familiar with the first pin "print" of LCD graphics?

They all display the following information. Serial printing displays through serial port monitor or Bluetooth; LCD displays through display screen.

The communication between the two modules also needs to use glue module to connect, and modify the pin number of ultrasonic graph, as shown in Fig. 4-2-23.

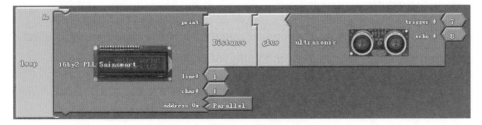

Fig. 4-2-23　Ultrasonic display program

Upload the program and observe the phenomenon.

(2) As shown in Fig. 4-2-24, when the range change is weak, the data displayed has little influence; but when the range change is large, it is not conducive to observation, and the data has a deviation of 10 times. What is the reason? The reason is that the data of the previous moment almost overlaps with the data of the next moment.

Fig. 4-2-24　Actual data

For example, the detection value at the previous moment is 122 cm, and the actual distance detected at the next moment is 15 cm. However, the LCD data will be 152. This is because the LCD did not eliminate the last digit 2, so there is an error.

When the frequency of change is too fast, the numbers will overlap. Because of the visual persistence of human eyes, people can hardly see the data clearly. Therefore, we need to find a solution.

(3) We need to erase the LCD, that is to erase the last data before displaying the new data.

There is exactly one such module in the "Generic Hardware" column, as shown in Fig. 4-2-25. There are many functions can be chosen.

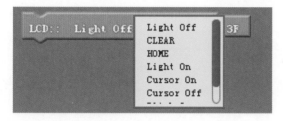

Fig. 4-2-25   Erase operation

We select "CLEAR" and change the communication mode to the same Parallel as 1602 LCD, as shown in Fig. 4-2-26.

Fig. 4-2-26   modification of communication mode

In order to make the data refresh slower, a delay function is added, as shown in Fig. 4-2-27.

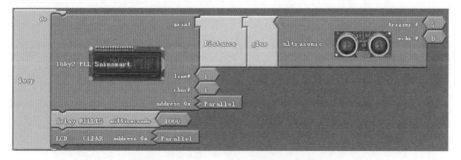

Fig 4-2-27   The main program

Note: The order of delay and CLEAR. If CLEAR is in front of delay, LCD will perform erase operation first, so that data can hardly be seen on LCD.

Experiment 2: Electronic fence.

Now that we know how ultrasonic measure distance, we will try to judge the results.

When the distance is greater than 15 cm, the tri-color LED is emit green; when the distance is less than or equal to 15 cm, the tri-color LED is emit red.

This is a conditional judgment, and there are only two cases, so select the "if/else" module, as shown in Fig. 4-2-28.

Fig. 4-2-28　"if / else" module

Comparing the ultrasonic result with 15, as shown in Fig. 4-2-29.

Fig. 4-2-29　Comparing the ultrasonic result with 15

Add the tri-color LED control program to the "if /else" module and connect the main function.

At this point, the electronic fence experiment is completed, as shown in Fig. 4-2-30.

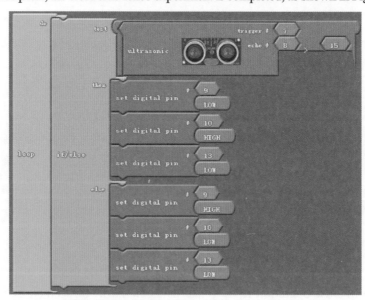

Fig. 4-2-30　Electronic fence experiment

Note: When downloading the program, be sure to press the white button. The order of CLEAR and delay when make the LCD display.

Expanded training: Can wireless communication be added to the program?

When there are many judgment conditions, such as 10, 15, 20, 30 and so on, is it appropriate to follow the above programming method? How to simplify? (Tips: The alarm information can be sent to the mobile phone through Bluetooth; the ultrasonic can be assigned as a variable, so that in the subsequent judgment, call the variable is OK.)

Subtask 2　Application of Door and Window Security System

1. Application scenarios

With the development of science and technology, people's quality of life has been greatly improved. People attach great importance to having a comfortable home after work. People's family anti-theft consciousness is also higher and higher. Doors and windows are essential in home, therefore, a higher security level of doors and windows security system is introduced.

2. Hardware introduction

One of the most widely used methods is called infrared detection. The infrared radiation sensor is installed on the door and window. When there is an obstacle invading, the target will be blocked and an alarm will be sent out.

1) Introduction of infrared

In addition to visible light, there are many invisible lights in nature, as shown in Fig. 4-2-31.

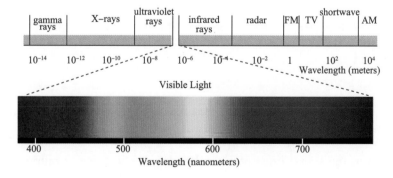

Fig. 4-2-31　Light types

Visible light refers to the light wave range visible to the naked eye, ranging from 400 nm (violet light) to 700 nm (red light), while the light with wavelength between 760 nm and 1 mm is called infrared ray, which is a kind of light that cannot be seen by the naked eye. With the help of some optical equipments, you can feel the infrared ray, and usually the infrared camera will convert it into visible green light. People's naked eyes will never see real infrared.

The infrared radiation alarm are installed on doors and windows. When the intruder passes through the infrared radiation sensor, the infrared radiation sensor will send out the alarm signal.

2) Module introduction

The infrared radiation sensor are composed of two parts, the infrared transmitter and the infrared receiver, as shown in Fig. 4-2-32. The transparent one is the transmitter and the black one is the receiver. The working principle is based on whether the receiver can receive the infrared radiation from the transmitter to output the signal, so as to realize the control of other circuits. The transmitter is only connected with GND and VCC, which plays the role of infrared emission; besides connecting GND and VCC, the receiver also needs to connect signal line, as shown in Table 4-2-4.

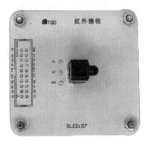

Fig. 4-2-32　Infrared radiation sensor

Table 4-2-4　Pins connection between IR sensor and core controller

No.	IR sensor pins	Core controller
1	VCC	VCC
2	GND	GND
3	OUT	D7

When the infrared light emitted by the infrared transmitter can be received by the receiver, the signal line outputs a low level; otherwise, it outputs a high level.

The buzzer is an integrated electronic buzzer, as shown in Fig. 4–2–33. It is widely used in computers, printers, copiers, alarms, electronic toys, automotive electronic equipment, telephones, timers and other electronic products.

Buzzers are divided into active buzzers and passive buzzers. The "source" here does not refer to the power supply, but to the oscillation source. That is to say, the active buzzer has its own oscillation source, so it will sound as soon as it is powered on, which is relatively simple to use; however, the passive buzzer has no internal oscillation source, so PWM square wave must be used to drive it, and different analog quantities can even make it emit different sounds tune.

Fig. 4–2–33　buzzer

In this experiment, the buzzer is used as the alarm device, so the active buzzer is OK. The pins connection between buzzer module and core controller is shown in Table 4–2–5.

Table 4–2–5　Pins connection between buzzer module and core controller

No.	Buzzer module pins	Core controller
1	VCC	VCC
2	GND	GND
3	IN	D6

3. Operation steps

Next, we through the door and window security system experiment to learn the application of infrared radiation sensor in intelligent security.

1) Flow chart

Before programming, first think about the overall process: Similar to the ultrasonic measuring experiment, when the infrared sensor completes the shooting, it outputs high level; when it is blocked, it outputs low level, which is used as the judgment condition to control the action of the buzzer, as shown in Fig. 4–2–34.

When the infrared sensors completes the shooting, the buzzer does not sound, and the Bluetooth APP receives the message "safe".

When the infrared sensors are blocked, the buzzer will sound and the Bluetooth APP will receive the message "warning"!

2) Graphic programming

Experiment 1: Output experiment of infrared radiation sensors.

Next, we through a small experiment to verify the output level of IR radiation and the previous introduction are the same.

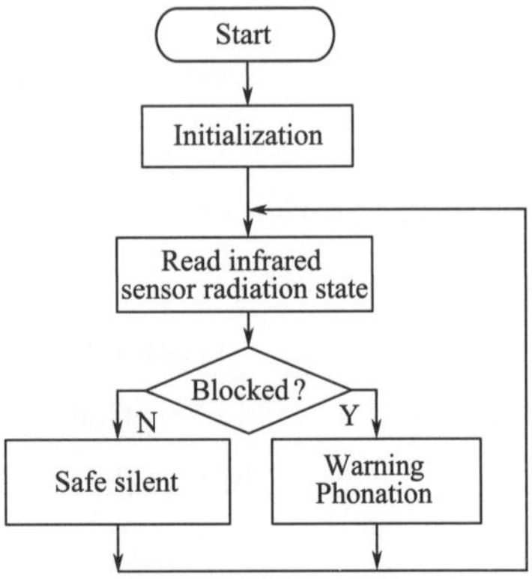

Fig. 4–2–34　Flow chart of Infrared radiation sensors

Selecting serial printing graphics, and display the level change of infrared radiation sensors through serial port, as shown in Fig. 4–2–35.

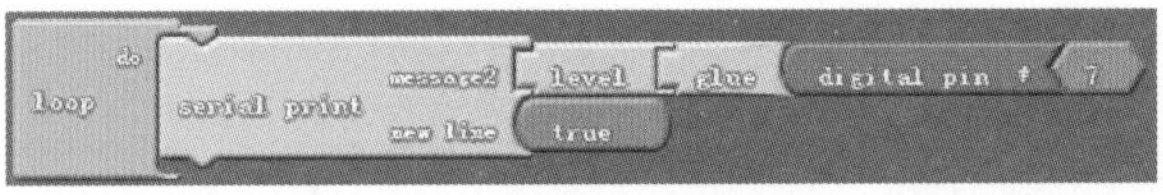

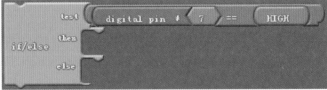

Fig. 4–2–35　Infrared radiation output 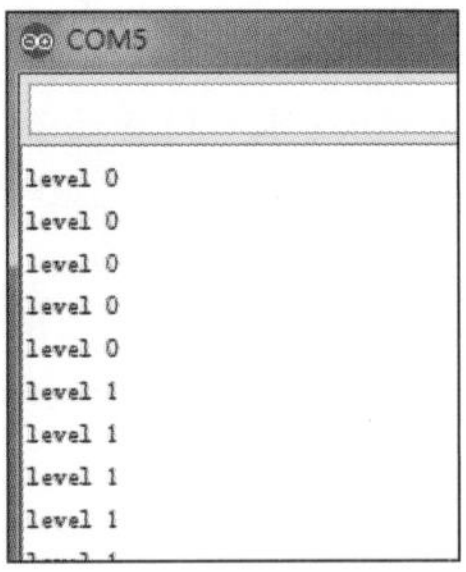　　　　Fig. 4–2–36　Output data

Experiment 2: Infrared radiation alarm experiment.

(1) We select the "if/else" graphic in the "control" column as the judgment condition, as shown in Fig. 4–2–37.

Fig. 4–2–37　"if / else"

(2) Adding the control program of buzzer, as shown in Fig. 4–2–38.

Fig. 4–2–38　security and alarm

(3) Finally, the combined program is shown in Fig. 4–2–39.

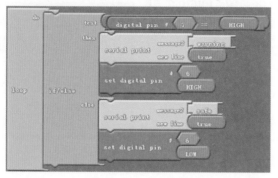

Fig. 4–2–39　Adding main program graphics

Note: when outputting other pin data by serial printing graphics, remember to use glue graphics for connection; the difference between "if" and "if/otherwise" graphics; when uploading the program, you must press the white button, and when using wireless Bluetooth communication, you must pop up the white button.

Expanded training: How to output the safe or warning information through display module? As shown in Fig. 4–2–40 and Fig. 4–2–41.

Fig. 4–2–40　Security status

Fig. 4–2–41　Alarm status

Task 3　Micromouse Controller Controls the Display System

With the rapid development of Internet technology, whenever the festival comes, the colorful lights and dazzling images at the gate of shopping malls and supermarkets will reflect each other. The multi-element aesthetic feeling greatly relaxes our body and mind and creates a strong festival atmosphere.

1. Application scenarios

In shopping malls and squares, we can see many screens, some display texts, some display patterns. Through close observation, it can be found that these screens are composed of a large number of small LEDs. By controlling the LEDs at different positions to turn on different colors, the text and pattern can be displayed, as shown in Fig. 4–2–42.

Fig. 4–2–42   LED lattice

The common display methods of LED display screen are column by column scanning (line by line scanning) and interlaced scanning (interlaced scanning). Column by column scanning (also known as non-interleaved scanning) is a method of encoding bitmap images. By scanning each column (row) of pixels, the video images are "drawn" on the screen. Interlaced scanning is similar to column by column scanning. The scanning mode is column by column. The odd column is scanned this time, and even column is scanned next time. In terms of clarity, column by column scanning is better, as shown in Fig. 4–2–43.

(a) Column by column scanning   (b) Interlaced scanning

Fig. 4–2–43   Different scanning modes

2. Module introduction

TQD lattice module is shown in Fig. 4–2–44, with common anode design and high-lighted red display; it is composed of 64 LEDs with 8×8.

According to the traditional LED control method, the whole 8×8 lattice needs 64 I/O ports. In order to save port resources, we specially set the lattice module. Each row and column share the same IO, so each LED corresponds to different rows and columns, and a total of 16 pins can realize data display.

When the row pin is low level and column pin is high level, the corresponding LED will be on.

The corresponding relationship between the lattice and the controller is shown in Fig. 4–2–45.

Fig. 4–2–44   Lattice module   Fig. 4–2–45   Pin diagram of lattice module

Among them, the analog ports A0-A3 can be directly used as digital ports D14-D17.

3. Operation steps

Next, we learn the method of screen display by scrolling the arrow figure.

Before programming, first think about the overall process: Row and column of lattice module share the same I/O, we select the column by column scanning for this experiment. Complete the static displays of arrows (8 kinds in total), as shown in Fig. 4–2–46. The scrolling of arrow can be realized by arranging them in order.

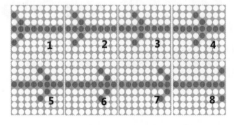

Fig. 4–2–46   Static displays of arrows

1) Static display and dynamic display flow charts

The flow charts of static display and dynamic display is shown in Fig. 4–2–47.

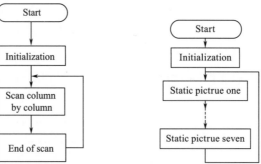

Fig. 4–2–47   Flow charts of static display and dynamic display

2）Graphic programming

Next, we will learn how to use the lattice module through a series of small experiments.

(1) Light up a single LED.

For example, row 4 (pin 5) column 1 (pin 17), as shown in Fig. 4–2–48.

Fig. 4–2–48   Light up a single LED

(2) Light up one row LEDs. For example, turn on all LEDs in the fourth line (pin 5), as shown in Fig. 4–2–49.

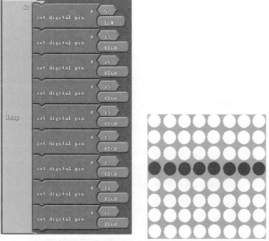

Fig. 4–2–49   Light up one row LEDs

(3) Light up one column LEDs. Turning on all LEDs in the second column (pin 16), as shown in Fig. 4–2–50.

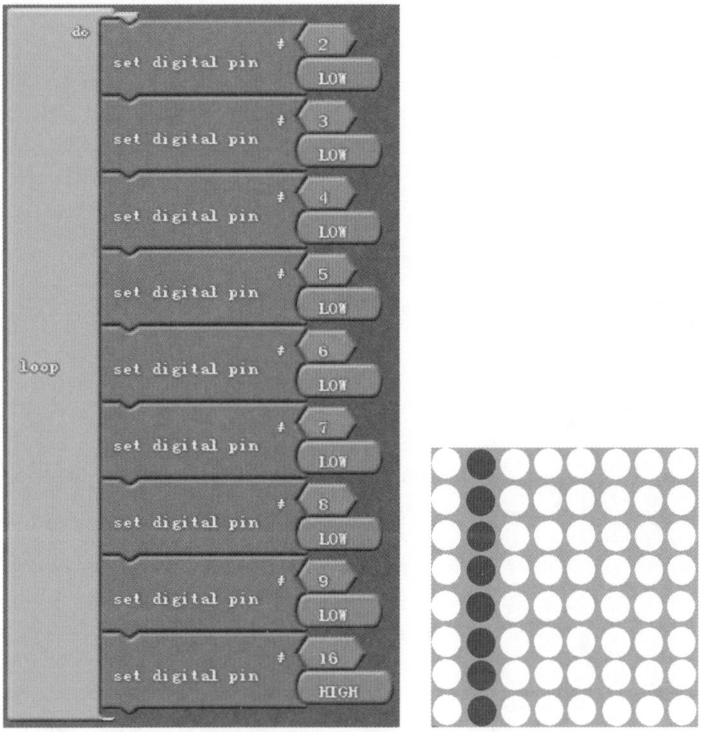

Fig. 4–2–50   Light up one column LEDs

(4) Lattice module reset. In different stages of scanning, different LEDs need to be turned on. In order to avoid graphic interference (similar to the clean screen function of LCD), it is necessary to reset the lattice module once before switching.

We adopt the method of level reverse setting to reset all 64 LEDs, that is, all rows are set to high level and columns are set to low level, as shown in Fig. 4–2–51.

Now that readers have learned how to use the lattice module, let's try to display the arrow.

(5) Static arrow. First, we light up the first static arrow in Fig. 4–2–46 and scan column by column.

Column 1, we light up the 2nd, 4th and 6th LED, as shown in Fig. 4–2–52.

Column 2, we light up the third to fifth LED, as shown in Fig. 4–2–53.

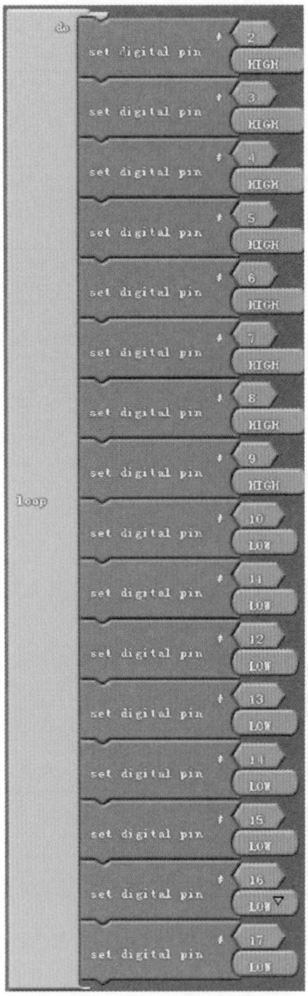

Fig. 4-2-51　Lattice module reset 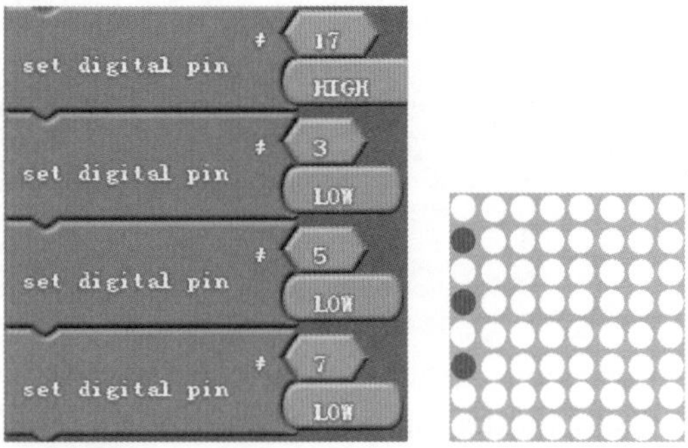　　Fig. 4-2-52　Lighting up the first column

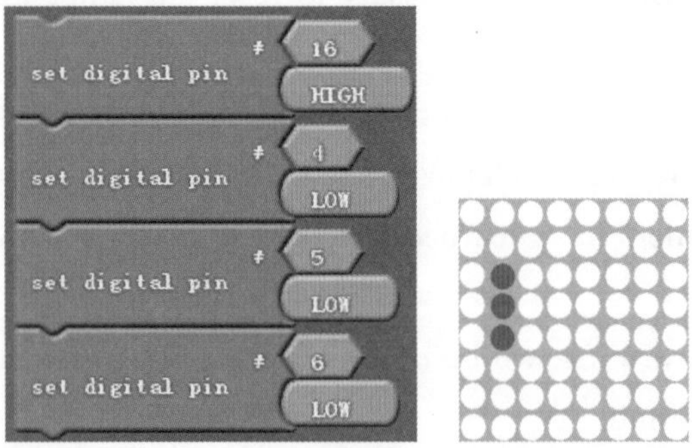

Fig. 4-2-53　Lighting up the second column

Column 3, we light up the fourth LED, as shown in Fig. 4-2-54.

Fig. 4–2–54　Lighting up the third column

The fourth LEDs are all turned on in the fourth to eighth columns, as shown in Fig. 4–2–55.

Reset must be added between column by column scanning, otherwise the lattice will be bright, as shown in Fig. 4–2–56.

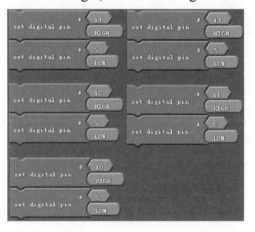

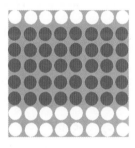

Fig. 4–2–55　Lighting up columns 4-8　　　Fig. 4–2–56　No reset

We name the reset subroutine "reset". The first static arrow program is shown in Fig. 4–2–57.

For the other seven static arrows, readers can try it out by modifying the eight columns LEDs.

(6) Arrow scrolling. After completing the eight static arrows above, you can try to roll the arrow.

We name them "one" "two" "three" … "seven" "eight".

① Now think about it. If you add sub functions to the main program, what will happen?

From the actual phenomenon, we can see that all the LEDs up again, and there are no dynamic effect, as shown in Fig. 4–2–58.

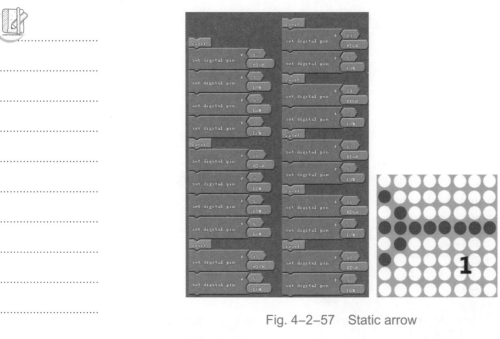

Fig. 4-2-57　Static arrow

What is the reason for this? Because the program runs very fast, reaching millisecond or even microsecond, human eyes cannot distinguish "one", "two", "three" … "seven" "eight" run alternately. Due to the visual persistence of the human eye, people's eyes superimpose the eight static arrows together, just as an image that superimposed by eight static arrows.

② Can we add delay between subroutines? As shown in Fig. 4-2-59.

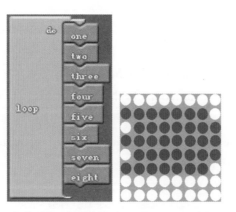

Fig. 4-2-58　Static arrow direct combination

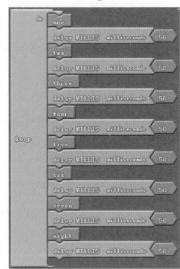

Fig. 4-2-59　Static arrow combination (adding 50ms delay)

Uploading program to observe the phenomenon, is it just the last column that glows? The LEDs brightness in other parts of the arrow is so low that it can hardly be observed.

Delay: keep the current output state of the controller unchanged and do no other operation until the end of the delay time, and then enter the next operation.

Because the last part of the sub function is the scanning of the eighth column, the duty cycle of the LED in other parts of the arrow is very low within 50 ms, and the brightness is almost invisible. Therefore, the program running result in Fig. 4-2-59 is like the light emission of the eighth column LED.

Other ways need to be tried to solve this problem.

③ The purpose of the delay function is to distinguish the eight static arrows. Is it possible to run "one" "two"..."seven" "eight" several times separately?

"one" "two"... "seven" "eight" run very fast in a single run, at the millisecond or even microsecond level. Then repeat each static graphic for 100 times or 200 times, is this visible to our naked eyes?

There are three kinds of cycle graphics in the "control" column, as shown in Fig. 4-2-60.

Fig. 4-2-60　Cycle graphics

The first is to check whether the test is true before running the loop. Otherwise, skip this program directly.

The second is to run the program once, and then verify whether the test is true. If it is true, it will loop, otherwise it will enter the next section of the program.

The third is unconditional cycle, forcing the cycle to set the number of times in advance, and then enter the next section of the program.

Here we can use the third unconditional cycle, and set the number of times to 200, as shown in Fig. 4-2-61.

Uploading program observation results, you can see the arrow srolling. Readers can try to change the number of cycles to change the speed of scrolling.

Note: When the arrow is displayed dynamically, the visual persistence feature of human eye is used. The size of the number of cycles determines the continuity of the scrolling, which can be tried many times until a suitable value is found.(no flicker should be observed by naked eyes)

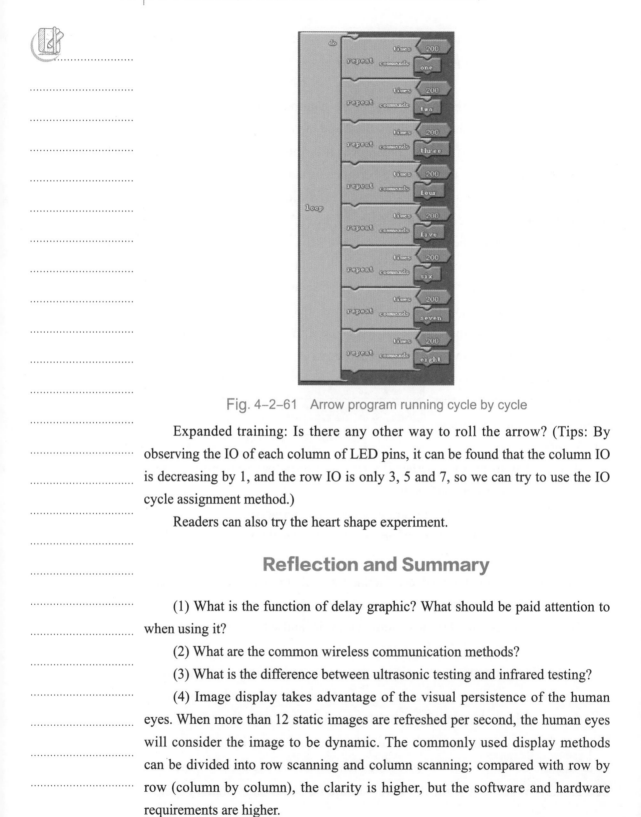

Fig. 4-2-61   Arrow program running cycle by cycle

Expanded training: Is there any other way to roll the arrow? (Tips: By observing the IO of each column of LED pins, it can be found that the column IO is decreasing by 1, and the row IO is only 3, 5 and 7, so we can try to use the IO cycle assignment method.)

Readers can also try the heart shape experiment.

Reflection and Summary

(1) What is the function of delay graphic? What should be paid attention to when using it?

(2) What are the common wireless communication methods?

(3) What is the difference between ultrasonic testing and infrared testing?

(4) Image display takes advantage of the visual persistence of the human eyes. When more than 12 static images are refreshed per second, the human eyes will consider the image to be dynamic. The commonly used display methods can be divided into row scanning and column scanning; compared with row by row (column by column), the clarity is higher, but the software and hardware requirements are higher.

Appendix

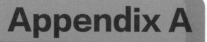

Appendix A

Micromouse Competition Going Popular in the World

Since 2019, it has been the most prosperous and fruitful period in the history of Micromouse competition.The International Micromouse Competition is held around the world, as shown in Fig.A–1.

In January, the International Micromouse Competition was held in Bombay, India.

In March, the APEC International Micromouse Competition was held in California, US.

In April, the International Micromouse Competition was held in Gondomar, Portugal.

In May, the IEEE Micromouse International Invitational Competition was held in Tianjin, China.

In June, the International Micromouse Competition was held in London, UK.

In August, the International Micromouse Competition was held in Chile.

In October, the International Micromouse Competition was held in Egypt.

In November, All Japan Micromouse International Competition was held in Tokyo, Japan.

Micromouse Competition Going Popular in the World

The International Micromouse Competition will become a booster of global higher education, vocational education, general education and technological innovation and integrated development of production and education. With the rapid development of artificial intelligence Micromouse competition, the education field timely introduced international well-known competitions to improve students' professional comprehensive ability, master the experience of practice and innovation, and help the integrated development of industry and education, and cultivate more excellent seed talents for the industry, profession and enterprise.

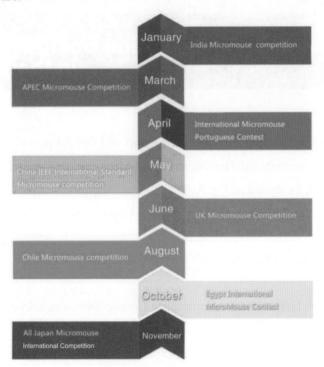

Fig. A–1　International Micromouse Competitions

1) China IEEE Micromouse International Invitational Competition

In 2009, Tianjin Qicheng Science and Technology Co., Ltd. introduced Micromouse competition into China, and carried out localized innovation and reform in the IEEE International Standard Micromouse Competition, which played a leading role in satisfying industrial optimization and upgrading, broadening international vision, gaining practice and innovation experience, and cultivating high-tech talents.

From 2016 to 2019, IEEE Micromouse International Invitational Competition has been successfully held for four times. The competition is hosted by Tianjin Municipal Education Commission and organized by Tianjin Qicheng Science and Technology Co. Ltd. and Tianjin Bohai Vocational Technical College, as shown in Fig. A–2.

Fig. A–2   The IEEE Micromouse International Invitational Competition in China
has been held since 2016

At present, IEEE Micromouse International Invitational Competition in China has set up "middle school, vocational college, bachelor's degree, master's degree and occupation" five competition group. It aims to improve the social participation and professional coverage of the competition. Micromouse has become an important carrier of systematic training and education. It fully embodies the combination of optical and electrical, software and hardware, control and machinery. While deducing the concept of "engineering" course, it extends and expands the concept of "innovative" course, which makes the content of students' learning and the teaching method of teachers have a new connotation, and truly focuses on the cultivation of comprehensive quality to create happy quality education.

IEEE Micromouse International Invitational Competition in China has the following characteristics.

(1) Participants: Facing not only college students, but also primary school, middle school and vocational workers, reflecting the characteristics of through training and lifelong education. It also includes international Micromouse professional players and previous international Micromouse competition winners.

(2) Maze site: There are 8×8 Micromouse maze sites for primary and secondary schools, and also have 16×16 full size classical Micromouse maze site for colleges and universities. Other than that, there is also a 25×32 half size Micromouse maze for elite players. Reflecting the extensibility of the competition, it takes the Micromouse competition as the core form, and students from different learning stages can participate in the competition.

(3) Competition events: There are not only Micromouse competition,

Video

China IEEE
Micromouse
International
Invitational
Competition

but also Robotracer race, which reflects the competition is both technical and engineering, and show the idea of engineering application oriented competition.

(4) Competition rules: A comparison of the similarities and differences in competition rules of general education, vocational education, higher education, vocational elite, as shown in Table A–1.

Table A–1　A comparison of similarities and differences in competition rules

Entry category	General education	Vocational education	Higher education	Vocational elite
Competition form	(1) Online debugging of APP. (2) Graphical programming. (3) Application of IOT intelligent sensing technology. (4) 8×8 maze race	(1) The oretical knowledge assessment. (2) According to the referee the task programming and implement the corresponding function (3) On-site technical defense. (4) 16×16 classical maze racing	(1) DIY appearance and structure mechanical design. (2) Hardware technology innovation. (3) Program algorithm innovation. (4) 16 × 16 classical maze racing	(1) DIY appearance and structure mechanical design. (2) Hardware technology innovation. (3) Program algorithm innovation. (4) 25×32 half size maze racing
Competition content	(1) Assembly task 10%. (2) Debugging task 40%. (3) Racing task 50%	(1) Theoretical assessment 20%. (2) Innovation 30%. (3) Speed race 50%	(1) Innovation 20%. (2) Speed race 80%	Speed race 100%

International Micromouse experts on-site training guidance is shown in Fig. A–3.

Fig. A–3　International Micromouse experts on-site training guidance

2) US APEC International Micromouse Competition

In 1977, the first exciting Micromouse competition was held in New York, US. It was co-sponsored by IEEE and APEC. Thus, the most influential international American APEC world Micromouse Competition was born. Known as one of the world's three major Micromouse competitions, it has held 34 competitions until 2019.

The official website of APEC: http://www.apec-conf.org/.

American Micromouse enthusiasts website: http://Micromouse usa.com/, as shown in Fig.A–4.

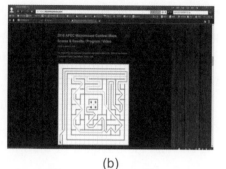

(a)   (b)

Fig. A–4   Screenshots of official website of American APEC International Micromouse Competition

Competition time: Between February and April every year.

Competition venue: Varies annually (previous venues include North Carolina, Texas, Florida, California, etc.). Every year, different countries such as the United States, the United Kingdom, Japan, South Korea, Singapore, India, China have actively participated in the competition, as shown in Fig. A–5.

Fig. A–5   The 30th American APEC International Micromouse Competition

3) UK International Micromouse Competition

Since 1980, UK International Micromouse Competition has become one of the internationally well-known Micromouse competitions.

Competition time: June every year.

Competition venue: Birmingham City University.

The competition is sponsored by UK Micromouse and Robotics Society. The

characteristic of the Micromouse competition in UK is that it does not restrict anyone to participate,whether you are from middle school, university or social personnel. All the players are divided into different groups, and the difficulty of maze is adjusted appropriately. The competition is divided into line follower, wall follower, maze solver and other projects, attracting more than 50 teams from more than 10 countries of the world.

UK Micromouse Competition scoring rules: In the 16×16 maze, the participating Micromouse need to complete the search from the beginning to the end and the traversal of the whole maze, solve the best route and complete the sprint from the beginning to the end. Scoring time = search time (time used to find the end for the first time) / 30 + sprint time (high-speed sprint with the shortest path from the starting point to the end) + penalty time (3s/time for knocking into the wall).

The offical website: https://ukmars.org/index.php/Main_Page, as shown in Fig. A–6.

Fig. A–6　Screenshot of official website of UK Micromouse Competition

4) All Japan Micromouse International Competition

All Japan Micromouse International Competition has been held 40 times from 1980 to 2019.

Competition time: November or December every year.

Competition venue: Tokyo, Japan.

The official Website: http://www.ntf.or.jp/mouse/Micromouse 2018/index.html, as shown in Fig. A–7.

Every year, Micromouse teams from more than 20 countries such as the United States, Britain, Japan, Singapore, China, Mongolia, Chile, Portugal participate in the competition, as shown in Fig. A–8.

● Video

All Japan
Micromouse
International
Competition

Fig. A–7   Screenshot of official website of All Japan Micromouse International Competition

Fig. A–8   Prize-giving of the 39th All Japan Micromouse International Competition

The competition consists of classic Micromouse event, half size Micromouse event and Robotrace event. The team consists of middle school students, college students and vocational elites. According to statistics, there are more than 300 teams. All Japan Micromouse International Competition can be said to represent today's international Micromouse technology field with the highest level and the strongest technology, so it has attracted much attention.

5) Chile Micromouse International Competition

The Ministry of Foreign Affairs of Chile hopes to promote the technological innovation of Chilean youth and international technological innovation, exchange and cooperation through Micromouse international competition, so as to promote the economic development of Chile. On December 3rd 2018, during the All Japan Micromouse International Competition, the Embassy of Chile in Japan hosted the "Chile International Micromouse Competition Seminar" and specially invited international experts (David Otten of the United States, Peter Harrison of the United Kingdom , Yukiko Nakagawa of Japan, Song Lihong of China, Benjamin of Chile, etc.) jointly discussed the unified standards and specifications for the Micromouse maze international competition in Chile, as shown in Fig. A–9 and Fig. A–10.

Fig. A–9　Meeting of ministry of foreign affairs, Chile—discussion on the development of Micromouse

Fig. A–10　Chile Micromouse international competition seminar

6) Portuguese Micromouse International Competition

Portuguese Micromouse International Competition has helded in Gondomar on April 27, 2019, sponsored by University of Tras-os-Montes and Alto Douro, Portugal and the Technical Executive Committee.

Portuguese Micromouse International Competition, which started in 2011, aims to provide a complete technological learning environment through the cultivation of creativity and ability, and has been successfully held for 9 times.

Competition time: Every April/May.

Competition venue: Portugal.

The official website: https://www.micromouse.utad.pt/, as shown in Fig. A–11.

On April 27, 2019, at 18:00 local time, in Gondomar Coliseumm, Portual, teams from UK, China, Portugal, Spain, Brazil, Singapore and other countries, were competing a tense international Micromouse maze competition. With

● Video

Micromouse
Portuguese
Contest

the steady search and fast sprint of Chinese Micromouse, applause and cheers thundered in Gondomar Coliseumm...Micromouse of Qicheng achieved breakthrough results and won the second place in the world, as shown in Fig. A–12.

Fig. A–11　Screenshot of official website of Portuguese contest Micromouse International Competiton

Fig. A–12　Micromouse of Qicheng won the world second place in the Portugal competition

　　Antonio Valente, chairman of Micromouse Portuguese Contest Organizing Committee, said after the contest that in recent years, China's comprehensive national strength and technical strength have been increasing, especially in the field of education, more and more attention has been paid to technological innovation and engineering literacy. TQD-Micromouse participated in the Portugal international competition for the first time, and its excellent results are very gratifying, as shown in Fig. A–13.

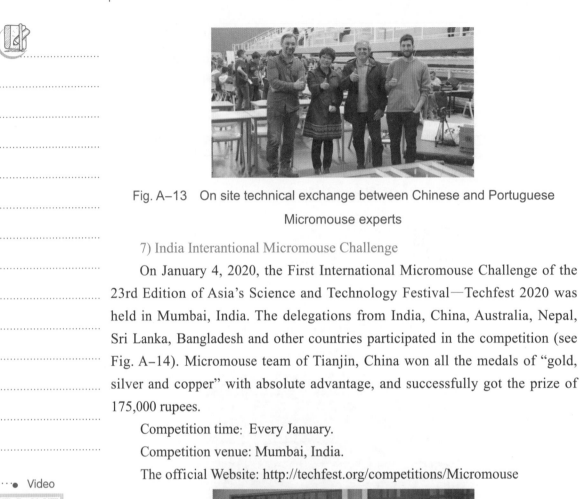

Fig. A–13   On site technical exchange between Chinese and Portuguese Micromouse experts

7) India Interantional Micromouse Challenge

On January 4, 2020, the First International Micromouse Challenge of the 23rd Edition of Asia's Science and Technology Festival—Techfest 2020 was held in Mumbai, India. The delegations from India, China, Australia, Nepal, Sri Lanka, Bangladesh and other countries participated in the competition (see Fig. A–14). Micromouse team of Tianjin, China won all the medals of "gold, silver and copper" with absolute advantage, and successfully got the prize of 175,000 rupees.

Competition time: Every January.

Competition venue: Mumbai, India.

The official Website: http://techfest.org/competitions/Micromouse

● Video

India
Interantional
Micromouse
Challenge

Fig. A–14   Group photo of International Micromouse Challenge, India

It is particularly worth mentioning that the Luban Workshop team of Chennai Institute of technology in India(see Fig. A–15) that adopted the IEEE

International standard equipment TQD-Micromouse-JD presented by China in 2017, won the champion of India domestic competition and the fourth place of the world elite group and won the prize of 5,000 rupees, becoming the star team of International Micromouse Challenge, India. Kasik, teacher of Indian Luban workshop, said that such excellent results of the team of Luban Workshop in Chennai Institute of Technology is the result of the joint efforts of teachers and students of Luban Workshop and the support of enterprises, Tianjin Qicheng Science and Technology Co., Ltd. in the past three years.

Fig. A–15　Group photo of winners in the International Micromouse Challenge, India

8) Egypt International Micromouse Competition

Egypt IEEE Institute of Electrical and Electronics Engineers has developed into one of the most influential international academic and technical organizations now. For more than 30 years, the institute has been promoting and guiding the development and innovation of power electronics technology. This technology includes the effective use of electronic components, the application of circuit theory and design technology, and the development of analysis tools for effective conversion, control and power conditions. Our members include outstanding researchers, practitioners and outstanding prize-winners.

As shown in Fig.A–16, IEEE publicizes Micromouse competition on the home page of its official website. IEEE Conference on Power Electronics and Renewable Energy offers generous prizes for the winners of a high-profile international Micromouse competition. Grand Prize: An equivalent of $1,000, Outstanding Performance Award: An equivalent of $700, Best Innovative Design Award: an equivalent of $500. The teams are open to students from Egypt or international engineering or related majors in Egypt, as well as high school students. Within each group, a maximum of two students are allowed.

Fig. A–16   Egypt International Micromouse Competition

The official website: http://www.ieee-cpere.org/International_Competition.html.

Photos of Egypt International Micromouse Competition as shown in Fig. A–17.

Fig. A–17   Photos of Egypt International Micromouse Competition

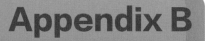

Appendix B

Entry-Level Micromouse Competition Analysis

IEEE International Invitational Competition is difficult, it is a challenging and interesting student competition that enjoys a certain reputation and influence at home and abroad. In terms of technology, the Micromouse competition covers Internet of things application technology, electronic information engineering technology, embedded technology, communication technology, software technology, computer network technology, information security technology, mobile communication technology, computer application technology, applied electronic technology, computer control technology, electromechanical integration technology, automation technology and other professional technologies. It involves skills and comprehensive professionalism in sensor detection, artificial intelligence, automatic control and electromechanical motion parts application and fully displaying the development of higher education and vocational education, and improves the training quality of high-quality and high skilled applied talents of electronic information.

In order to further deepen the education reform of ordinary middle schools, strengthen the integrated development of vocational education and general education, and explore diversified education modes, Tianjin Education Commission, from 2016 to 2019, held four consecutive sessions of Tianjin secondary vocational college students' skills competition, integration of general education and vocational education Micromouse competition, which integrates mechanical, electronic, optical, automatic control, artificial intelligence and other multi-disciplinary integration technologies. middle school students should be popularized to build a bridge between general education and vocational education. The following is a more representative of the general vocational integration of Micromouse competitions as an examples to do the analysis.

1) Analysis of the secondary vocational school integration of general education and vocational education Micromouse competition in 2017

The competition team form is very innovative, each competition team is composed of two secondary vocational students and one middle school student. The competition content includes theoretical knowledge assessment and practical operation. The middle school student and secondary vocational school students work together to complete the competition task, as shown in Fig. B–1 and Fig. B–2.

Fig.B–1　The referee assesses the actual operation tasks of the participating

Fig.B–2　The referee assesses the theoretical knowledge of the team

The 8×8 maze chart used in this competition is very classic. It not only shows the high-speed movement performance of Micromouse, but also shows the precise control of Micromouse through multiple continuous turns, as shown in Fig. B–3 and Fig. B–4.

Crucial point A: To investigate the students' ability of sensor detection and motor operation parameters debugging. Micromouse makes 12 consecutive turns in this crucial point. Because there is no straight way to correct the posture, there is a very high requirement for the accuracy of the operating parameters of Micromouse.

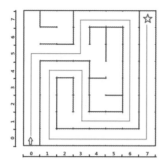

Fig.B-3   The optimal path

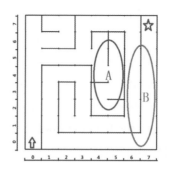

Fig.B-4   Analysis of the maze

Crucial point B: Because it is the first Micromouse competition for middle school students, four long straight paths are designed on the maze, which not only effectively reduces the overall difficulty of the maze, but also reflects the speed advantage of Micromouse.

2) Analysis of the secondary vocational school Jing-Jin-Ji Micromouse competition in 2018

In order to further implement the coordinated development of Jing-Jin-Ji education, schools from Beijing, Tianjin and Hebei participated in the competition. Students use sensing, mechanical and electrical, automatic control and other knowledge and skills to complete the assembly, debugging, innovation and racing tasks (see Figure B–5).

Fig.B-5   The Jing-Jin-Ji Micromouse competition in 2018

The 8×8 maze used in this competition is very helpful for the basic learners of popular science education to learn the entry-level intelligent algorithm, as shown in Fig. B–6 and Fig. B–7.

Crucial point A: This competition maze for the Micromouse algorithm, designed the left hand rule and the right hand rule examination. There are intersections on both sides of the crucial point A. Fully investigate the students' ability to master the basic intelligent algorithm, because each turn selection is very important. If the choice is wrong, it will affect whether Micromouse can find the end point smoothly.

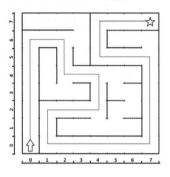

Fig.B–6   The optimal path   Fig.B–7   Analysis of the maze

Crucial point B: The maze design of this area has a high degree of openness. In the process of Micromouse search, the examination of 6 times of continuous turning without correction plays an important role in cultivating students' intelligent basic algorithm learning.

3) Analysis of the secondary vocational school vocational skills Micromouse competition in 2019

Students from 17 vocational colleges and middle schools in Tianjin participated in the competition (see Fig.B–8). The teachers and students from Honghe Hani and Yi Autonomous Prefecture, Nujiang Lisu Autonomous Prefecture and Hotan, Xinjiang Province were particularly attractive to watch the competition. In the era of rapid development of science and technology, intelligent control, as a subversive technology affecting the development of all aspects of society, has a significant impact on students' life and learning. These students learn to observe the Micromouse competition for the first time, which is very helpful for them to enhance their awareness of scientific and technological innovation and practical design ability. Help, looking forward to the next competition, students from Xinjiang and Western Yunnan can also compete in the same competition, to promote teaching and learning by competition, and strive to walk out of the poverty alleviation and vocational education.

Fig.B–8   The field programming task

The 8×8 maze used in this competition highlights the ornamental and learning features in the design of the maze, considering the viewing effect of the students. In addition to the assessment of basic intelligent algorithms, i.e. the left-hand rule and the right-hand rule. The more difficult "ring" maze path is designed, as shown in Fig.B–9 and Fig.B–10.

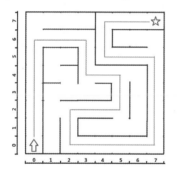 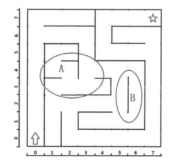

Fig.B–9   The optimal path   Fig.B–10   Analysis of the maze

Crucial point A: This area is one of the paths for Micromouse to reach the destination. Three consecutive turns have put forward higher requirements on the sensor detection accuracy and motor operation stability of Micromouse. Because this route is short, Micromouse, which generally adopt advanced algorithms, usually choose this route.

Crucial point B: This area is another path to the destination. In the maze design, in accordance with the principle of increasing difficulty year by year, a "ring" trap path is specially designed. For beginners, if the level of intelligent algorithm is not high enough, it will be difficult to pass the road smoothly.

4) Practical training example of entry level Micromouse maze (see Fig.B-11 to Fig.B-13)

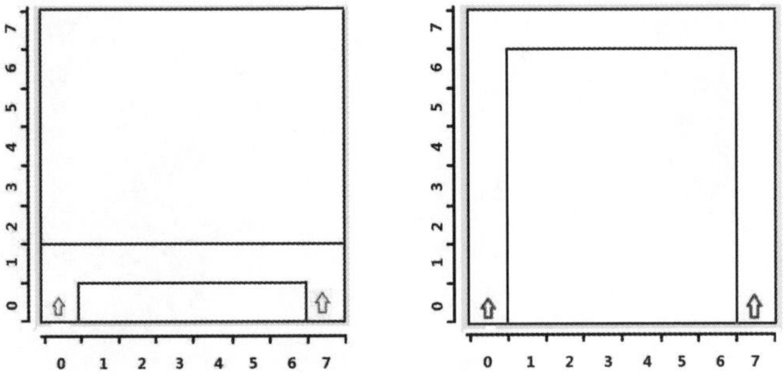

Fig.B-11　The primary practice maze

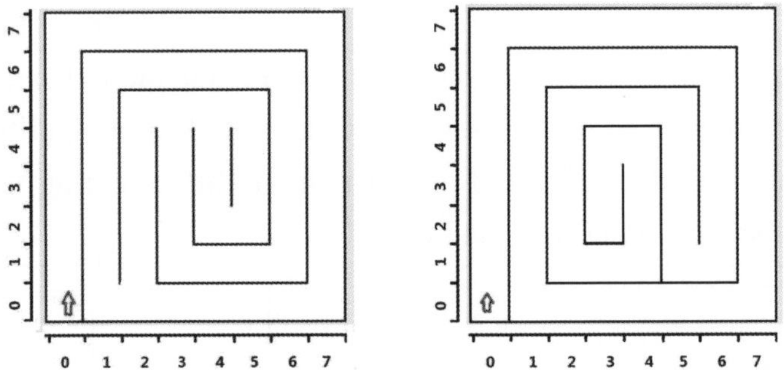

Fig.B-12　The ability improvement maze

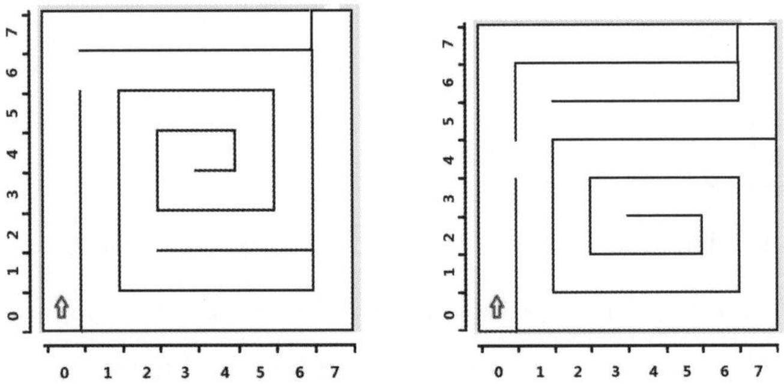

Fig.B-13　The advanced competition maze

Appendix C

Device List of TQD-Micromouse-JQ

Device list of TQD-Micromous-JQ is shown in Tabel C–1.

Table C–1　Device list of TQD-Micromouse-JQ

No.	Name	Quantity	Remarks
1	TQD-Micromouse-JQ	1	
2	Charger	1	
3	USB line	1	
4	Battery	1	
5	Disk	1	

Appendix D

Device List of TQD-IOT Engineering Innovation Course Platform

Device list of TQD-IOT engineering innovation course platform is shown in Tabel D–1.

Table D–1　Device list of TQD-IOT engineering innovation course platform

No.	Name	Quantity	Remarks	No.	Name	Quantity	Remarks
1	TQD-IOT platform	1		13	Passive buzzer	1	
2	Control module	1		14	IR sensor	1	
3	Bluetooth module	1		15	Human body IR sensor	1	
4	LCD module	1		16	Vibration sensor	1	
5	Single-color LED I	1		17	Fan module	1	
6	Single-color LED II	1		18	Temperature sensor	1	
7	Key module	1		19	Flowing water lights module	1	
8	Tri-color LED	1		20	Lattice module	1	
9	Photosensitive sensor	2		21	MP3 player module	1	
10	Potentiometer sensor	1		22	Voice control module	1	
11	Ultrasonic sensor	1		23	Water module	1	
12	Active buzzer	1		24	Servo module	1	

Appendix E

Teaching Content and Class Arrangement

The reference teaching hours are 60, and the allocation is shown in Table E–1.

Table E–1　Teaching content and class arrangement

No.		Teaching Contents	Teaching hours
Chapter 1　Elementary Knowledge	Project 1　Evolution of Micromouse		24
	Project 2　Micromouse Hardware Structure		
	Project 3　Development Environment of Micromouse		
	Project 4　Basic Function Control of Micromouse		
Chapter 2　Comprehensive Practice	Project 1　Interaction Control of Micromouse		14
	Project 2　Attitude Control of Micromouse		
Chapter 3　Advanced Skills and Competitions	Project 1　Analysis of Common Algorithms		12
	Project 2　Advanced Control Function of Micromouse		
Chapter 4　Extended Application	Project 1　Structural Composition of TQD-IOT Engineering Innovation Course Platform		10
	Project 2　IOT Extended Application of Micromouse Technology		
Total			60

Appendix F

The Circuit Diagram Symbol Comparison Table

The circuit diagram symbol comparison table is shown in Table F–1.

Table F–1　The circuit diagram symbol comparison

No.	Name	Drawing methods under China national standard	Drawing methods in software
1	Light-emitting diode		
2	Resistor		
3	Adjustable resistor component		

Appendix G

Bilingual Comparison Table of Glossary

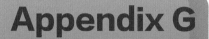

Bilingual comparison table of glossary as shown in Table G–1 to Table G–3.

Table G–1 Related to Micromouse

English	Chinese	English	Chinese
main control module	核心控制模块	PWM signal generator driver module	PWM信号发生器模块
main control chip	主控芯片	the front-right and thefront-left, the left, the front and the right	5组红外传感器方向
input module	输入模块	The g segment	g段
output module	输出模块	coreless DC motor	空心杯直流电动机
main control circuit	核心板电路	stepping motor	步进电动机
power circuit	电源电路	motor drive circuit	电动机驱动电路
control circuit	控制电路	truth table	真值表
peripheral circuit	外围电路	H-bridge circuit	H桥电路
Keyboard-display circuit	键盘显示电路	rotate	转动（步进电动机）
JTAG interface circuit	JTAG接口电路	electronic components	电子元器件
key-pressing circuit	按键电路	crystal oscillator	晶振
data transmission	数据传输	capacitance	电容
human-computer interaction system	人机交互系统	adjustable current-limiting resistance	限流可调电阻
suction fan technology	吸地风扇技术	Digitron	数码管
duty cycle	占空比	peripheral devices	外围器件
angular velocity	角速度	pulse oscillation circuit	脉冲振荡电路
IRsensor	红外线传感器	pulse signal	脉冲信号
infrared detection circuit	红外检测电路	square wave	方波
infrared light	红外线	perceptual system	感知系统
infrared calibration	红外校准	carrier frequency	载波频率
infrared intensity	红外强度	schematic Diagram	原理图
infrared transmitter	红外发射头	software interface	软件界面
infrared receiver	红外接收头	driver library	驱动库

Table G–2　Related to competition

English	Chinese
cell	单元格
wall	挡板
post	立柱
competition maze	竞赛场地
the start	起点
the destination	目的地/终点
the coordinate in the maze	迷宫坐标
crossing	路口
electronic automatic scoring system	电子自动计分系统
competitor	参赛队员
micromouse competition	智能鼠竞速比赛
the optimal path	最优路径
trajectory	轨迹
passage way	通道

Table G–3　Related to intelligence algorithm

English	Chinese	English	Chinese
the bottom driver program	底层驱动	differential-speed control	差速控制
the top algorithm program	顶层算法	straight movement	直线运动
algorithm	算法	turning	转弯
strategy	策略	correct the attitude	校正车姿
rule	法则（左、右手法则）	attitude correction	运行校正
the right-hand rule, the left-hand rule, the central rule	右手、左手、中心法则	core function	核心函数
90-degree turning/180-degree turning	90°、180° 转弯	time sequence status	（驱动步进电动机的）时序状态
programming and realizing	编程并实现	moving forward one cell	前进一格
step map	等高图	waiting for button press	按键等待
cycle detection	循环检测	determining the attitude	判断车姿
movement control in picture-8-shaped path	"8字形"路径运行控制	waiting one step	暂停一步
obstacle avoidance	实现避障	accurate turning control	精确转弯控制
motion attitude control	运动姿态的控制	closed-loop control	闭环控制
two-wheel difference speed	两轮差速	absolute direction	绝对方向
path planning and decision algorithm	路径规划和决策算法	relative direction	相对方向
struct	结构体		

Appendix H
The International Curriculum Standard for "Micromouse Design Principles and Production Process"

(Applicable to training courses in secondary vocational colleges)

1. Course information

Applicable objects: Three year secondary vocational education level students.

Applicable majors: Electromechanical technology application (051300), electronic technology application (091300), computer application (090100), computer network technology (090500), electrical operation and control (053000).

Course type: Theory and practice.

Suggested semester: The third semester.

It is suggested to set up leading courses, like electrical and electronic technology and application, mechanical drawing/CAD, microcomputer theory and interface technology, electronic measurement technology and instrument, C language programming, single chip microcomputer technology application.

It is suggested that follow-up courses should be offered like, sensor application technology, motor technology, single chip microcomputer principle and application, automatic control principle, electronic whole machine assembly and debugging.

2. Course characteristics

1) In line with the *National Standard of Undergraduate Professional Teaching Quality of General Institutions of Secondary Learning* issued by the Ministry of Education

Micromouse Design Principles and Production Process covers many specialties, such as computer application, electronic and information technology, electronic technology application, electromechanical technology application, electrical operation and control, etc., involving the skills and comprehensive professional quality of sensor detection, artificial intelligence, automatic control and electromechanical moving parts application. The curriculum design and implementation adopt the method of theory and practice integration, combine brain and hand, and run through technical skills training as a whole, so as to serve the internalization of professional ethics.

2) Connecting with the international teaching concept and serving the cultivation of skilled talents

On the basis of conforming to Chinese standards, the course content is connected with IEEE international competition content, and the design concept of "application, project and task" is embodied in the curriculum design of teaching materials. The internationally popular ideas of task orientation, systematization of working process and innovation of engineering practice are adopted to enable students to learn professional knowledge, and at the same time, communicate and cooperate with each other stage goals and other forms to improve the comprehensive quality of students.

3) Serve "the Belt and Road" initiative, spread China's educational standards

The content of this course is highly integrated with the "Luban Workshop" construction projects in many countries. As China's new name card, this course serves "the Belt and Road" initiative, which provides China's educational standards and provides rich practical teaching resources for all countries along "the Belt and Road" route, serving the training of skilled personnel in various fields.

3. Course objectives

This course aims to cultivate students' abilities of Micromouse system structure analysis and assembly, integrated development environment construction, sensing and detection signal debugging, Micromouse straight and turning control and posture correction, Micromouse intelligent search and path planning, inquiry ability and ability to solve practical problems with theoretical knowledge.

1) Knowledge objectives

(1) Master the basic hardware structure of Micromouse, and understand the basic principles and methods of assembly and debugging.

(2) Understand the control principle of the core control board of Micromouse.

(3) Understand the human-computer interaction control system of Micromouse.

(4) Master the common algorithm design of Micromouse, and understand the related knowledge of intelligent search and path planning.

(5) Understand the principle of Micromouse multi-sensor cooperation.

2) Capability objectives

(1) Be able to use common instruments correctly.

(2) Have the ability to consult the manual, read the product manual and other materials.

(3) Have the ability to analyze and remove the common faults in the debugging process of Micromouse equipment.

(4) Have the ability to explore and use theoretical knowledge to solve practical problems.

(5) Master the basic debugging method of Micromouse sensor and motor.

(6) Master the development environment and basic use method of Micromouse.

3) Quality objectives

(1) Have the ability of communication and teamwork.

(2) Have the work style of innovation, dedication and joy, and the craftsman spirit of preciseness, refinement and realism.

(3) Have the awareness of safety, quality and responsibility.

4. Teaching contents and requirements

Recommended total class hours: 60 class hours.

Teaching contents and requirements as shown in Table H–1.

Table H–1   Teaching contents and requirements

No.	Contents	Teaching requirements	Recommended class hours	Key and difficult points
1	Chapter 1 Project 1 Evolution of Micromouse	Understanding the origin of Micromouse. Be familiar with the competition and debugging environment of Micromouse	4	Debugging environment of Micromouse

Continued

No.	Contents	Teaching requirements	Recommended class hours	Key and difficult points
2	Chapter 1　Project 2 Micromouse Hardware Structure	Master the hardware structure of Micromouse and the relationship between each part. Understand the selection principle and method of key components of Micromouse. Deeply understand the working principle of Micromouses' core circuit	8	Micromouses' core circuit working principle
3	Chapter 1　Project 3 Development Environment of Micromouse	Familiar with Arduino development environment. Master the program download method of Micromouse	4	Familiar with Arduino development environment
4	Chapter 1　Project 4 Basic Function Control of Micromouse	Let the eyes of Micromouse see the world, and master the basic regulation of Micromouse sensor. Let the Micromouses' legs move, master the basic adjustment of Micromoouse motor	8	Sensor detection and motor drive
5	Chapter 2　Project 1 Interaction Control of Micromouse	Understand the operation control of wireless Bluetooth and master the cooperation of Micromouses' eyes and legs	8	Cooperation between eyes and legs
6	Chapter 2　Project 2 Attitude Control of Micromouse	Let Micromouse run and learn how to turn. Master Micromouse motor control principle and drive method, complete straight line and turn control. Master how to correct the action posture of Micromouse through sensor	8	Straight and turning control
7	Chapter 3　Project 1 Analysis of Common Algorithms	Master the left-hand and the right-hand rule	4	Principle of maze search optimization rule
8	Chapter 3　Project 2 Advanced Control Function of Micromouse	Master the principle and implementation of multi-sensor cooperative work, multi-thread working principle and implementation, intelligent mouse to avoid obstacles and run flexibly	8	Implement obstacle avoidance
9	Chapter 4　Project 1 Structural Composition of TQD-IOT Engineering Innovation Course Platform	Understand the relationship between TQD-IOT platform and Micromouse. 　Understand the hardware composition of TQD-IOT platform	2	Differences between usage methods of TQD-IOT platform and Micromouse
10	Chapter 4　Project 2 IOT Extended Application of Micromouse Technology	Master the control method of Micromouse controller to control the lighting system, the security system and the display system.	6	Complete different IOT experiments based on different applications

5. Mode and method

(1) Requirements of teaching mode: The project-based teaching mode should be adopted. Under the engineering background, the engineering practice oriented and real task driven teaching should be implemented. The practice carrier Micromouse and situational IEEE labyrinth should be used in the teaching, which is the real scene, the combination of virtual and real, and the combination of software and hardware. The process of practice requires that "knowledge and technology advance together, morality and technology should be combined". Guide students to use the knowledge they have learned to improve their ability to analyze and solve problems independently.

(2) Specific requirements for teachers: Bachelor degree or above, intermediate title or above, teaching experience in information technology and processing and manufacturing specialty in vocational colleges.

(3) Practical teaching environment conditions: The training site should be flat, bright and well ventilated, with an area of no less than 100 m^2. It should be equipped with training station, training computer, Micromouse platform, IEEE competition maze, electronic information equipment platform for expansion, etc., and the number can meet the training teaching needs of 40 students in a standard class.

(4) Requirements for the compilation of teaching tasks: The project assignment book should specify the contents of the teachers' lecture (or demonstration); the requirements of learners' preview; the overall arrangement of the project and the training time and content of each project should be put forward. The requirements for group arrangement and group discussion (or operation) should also be specified clearly.

6. Assessment and performance evaluation methods

(1) Assessment method: Examination.

(2) Performance evaluation: Theoretical and practical assessment.

The total score is calculated on the 100 point system, which is divided into two parts: The usual score assessment and the final comprehensive assessment. Generally speaking, attendance, project assignment and test account for 30% of the total score in the usual performance assessment, while the comprehensive assessment at the end of the term accounts for 70% of the total score in the theoretical examination and practical operation assessment, of which the theoretical examination accounts for 30% and the practical operation assessment accounts for 40%.

7. Teaching materials

1) Recommended teaching materials

王超，高艺，宋立红. 智能鼠原理与制作：基础篇[M]. 北京：中国铁道出版社有限公司，2019.

2) Reference materials

[1] 樊胜民，樊攀，张淑慧. Arduino编程与硬件实现[M]. 北京：化学工业出版社，2019.

[2] 陈吕洲. Arduino程序设计基础[M]. 北京：北京航空航天大学出版社，2015.

8. Prepare and review

Prepared by: Li Jun, Wang Xiaoqin, Song Lihong, Feng Qiang, Zhang Jiaquan, Chen Likao, Qiu Jianguo

Reviewed by: Gong Wei, Gao Yi, He Dongmei

July 16, 2020

智能微型运动装置（Micromouse）技术与应用系列丛书

天津市"一带一路"联合实验室（研究中心）项目研究成果

工程实践创新项目（EPIP）教学模式规划教材

智能鼠原理与制作

（初级篇）

王 超 高 艺 宋立红 编著

王晓芹 严靖怡 周繁雨 编译

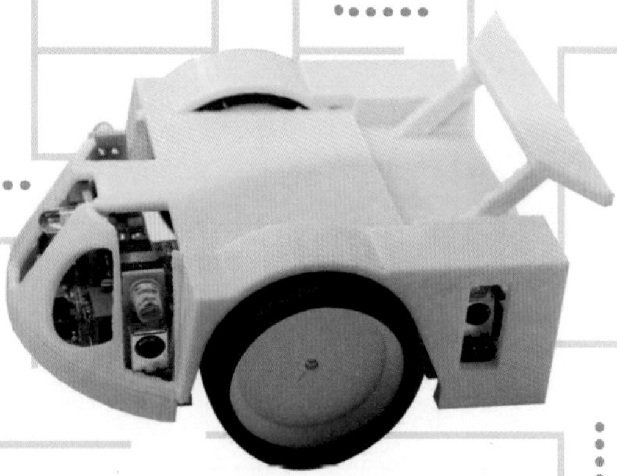

中国铁道出版社有限公司

CHINA RAILWAY PUBLISHING HOUSE CO., LTD.

内 容 简 介

本书为中英双语版，以天津启诚伟业科技有限公司提供的 TQD-Micromouse-JQ 智能鼠为载体，是智能微型运动装置（Micromouse）技术与应用系列丛书的初级篇。

本书以真实工程项目为背景，通过"基础知识篇"、"综合实践篇"、"竞赛创新篇"和"拓展应用篇"四篇讲述了智能鼠的发展、硬件、开发环境、功能调试；智能鼠的交互控制、姿态控制；智能鼠算法解析、高级功能；智能鼠技术应用扩展等。同时，本书附录提供了国际 Micromouse 走迷宫竞赛相关知识、智能鼠器件清单、智能鼠迷宫图库、专业词汇中英文对照表、国际实训课程标准等丰富资源。

本书在重要的知识点、能力点和素养点上，配有丰富的视频、图片、文本等资源，学习者可以通过扫描书中二维码获取相关信息。

本书适合作为基础教育学校开展职业启蒙、科技活动和特色教育的教材，也可作为职业教育的教材，还可作为相关工程技术人员培训用书及智能鼠爱好者参考用书。

图书在版编目（CIP）数据

智能鼠原理与制作 . 初级篇：汉、英／王超，高艺，
宋立红编著 . —北京：中国铁道出版社有限公司, 2021. 1
（智能微型运动装置（Micromouse）技术与应用系列丛书）
ISBN 978-7-113-27522-8

Ⅰ. ①智… Ⅱ. ①王… ②高… ③宋… Ⅲ. ①智能机
器人－程序设计 Ⅳ. ① TP242. 6

中国版本图书馆 CIP 数据核字（2020）第 273201 号

书　　名：**智能鼠原理与制作（初级篇）**
作　　者：王　超　高　艺　宋立红

策　　划：何红艳　　　　　　　　　　编辑部电话：（010）83552550
责任编辑：何红艳　绳　超
封面设计：刘　颖
责任校对：孙　玫
责任印制：樊启鹏

出版发行：中国铁道出版社有限公司（100054，北京市西城区右安门西街 8 号）
网　　址：http://www.tdpress.com/51eds/
印　　刷：北京柏力行彩印有限公司
版　　次：2021 年 1 月第 1 版　2021 年 1 月第 1 次印刷
开　　本：787 mm×1 092 mm 1/16　印张：19　字数：318 千
书　　号：ISBN 978-7-113-27522-8
定　　价：66. 00 元

作者简介

王 超

天津大学电气自动化与信息工程学院教授，教育部高等学校自动化类专业教学指导委员会委员，主要从事多相流检测与仪器和电学层析成像的研究（ERT, ECT, EMT 和 EST），在天津大学教授计算控制技术和工业控制网络课程，自 2010 年起，首次将智能鼠作为重要的实践教学载体引入电气自动化与信息工程学院。2018 年，在第 33 届 APEC 国际电脑鼠竞赛中，天津大学的两支队伍包揽冠亚军。

高 艺

南开大学电子信息工程学院硕士生导师，电子信息实验教学中心副主任，天津市单片机学会青年骨干工作委员会副主任，多项天津市大学生竞赛及职业技能竞赛裁判组成员。先后参与多项"国家高技术研究发展计划（863 计划）项目"、"天津市科技支撑计划重点项目"以及横向科研项目。多次作为指导教师带队参加全国大学生电子设计竞赛、天津市电子设计竞赛、天津市物联网竞赛、天津市大学生 IEEE 电脑鼠竞赛、APEC 国际电脑鼠竞赛、全国机器人大赛等。

宋立红

天津启诚伟业科技有限公司总经理，启诚智能鼠创始人。多年来专注于高等教育、职业教育、基础教育领域的嵌入式、物联网、人工智能等教学仪器设备研发、设计、生产、推广、服务工作。40 余次赞助支持大学生学科竞赛"启诚杯"智能鼠走迷宫赛项及职业院校技能竞赛智能微型运动装置（智能鼠）赛项等。从 2016 年开始，积极致力于国际项目鲁班工坊技术支持服务工作，启诚智能鼠作为中国创新型教育装备，伴随鲁班工坊不远万里前往泰国、印度、印尼、巴基斯坦、柬埔寨、尼日利亚、埃及等国家，受到所在国师生一致青睐，为"一带一路"倡议做出了努力和贡献。

王晓芹

天津市东丽区职业教育中心学校一级教师，加工制造系副主任，区级学科带头人。2008 年荣获天津市中职院校教师技能大赛一等奖；2014 年荣获天津市教育教学成果二等奖；2016 年荣获天津市信息化教学能力大赛一等奖、全国说课大赛二等奖；2016 年第二届微课大赛全国一等奖；多次作为指导教师带队参加天津市中职生电子电路调试技能比赛、IEEE 电脑鼠竞赛、APEC 国际电脑鼠竞赛等。

严靖怡

天津启诚伟业科技有限公司总经理助理，就读于美国加州大学圣克鲁兹分校。2015 年担任美国 APEC 智能鼠国际竞赛组委会主席 MIT David Otten 教授在中国交流访问期间随行翻译。2018 年担任新蒙古教育集团董事会副主席 Davaanyam 访问天津智能鼠竞赛交流活动随行英语翻译及会议同声传译。2018 年自费到柬埔寨做志愿者，担任柬埔寨国

立理工学院 Bun Phearin 校长鲁班工坊智能鼠项目翻译，帮助柬埔寨学生学习中国智能鼠技术，为"一带一路"沿线国家教育发展做出了努力和贡献。

周蘩雨

美国中央俄克拉荷马大学英语教育学硕士，现任职于天津机电职业技术学院国际交流与合作处。在美期间参与了中英语言

对比研究等研究项目，在孔子课堂教授了大批以英语为第二语言的非美籍学生，从中获取了大量教学经验。

2018 年参加夏季达沃斯论坛，为国际能源论坛秘书长孙贤胜做专属随行翻译及事务官。2018 年参与葡萄牙鲁班工坊创设工作；2018 年发表《"鲁班工坊"教学模式在电脑鼠实践教学中的应用》中英双语论文。2019 年参与马达加斯加鲁班工坊创设工作，其间完成了多场会议翻译、随行翻译及笔译，累计笔译文件及材料超过 20 万字。

I am very pleased to find out that you are going to write a book about micromouse. This contest is a fantastic way to learn about electromechanical systems and integrating hardware and software. I have learned a great deal during my 30 years with this contest and I am sure your readers will also. Congratulations and good luck with this endeavor.

David Otten
APEC micromouse chair

我很高兴得知你们要编写关于智能鼠的书。智能鼠大赛是学习机电系统和集成软硬件的绝佳方式。在过去经历的30年比赛中，我学到了很多东西，我相信你们的读者也会这样。祝贺你们并祝好运！

Mr. David Michael Otten

美国麻省理工学院高级研发工程师。

国际智能鼠走迷宫教育教学专家。

多年从事AI机器人开发研究工作，连续30余届美国APEC国际电脑鼠竞赛组委会主席，曾多次参加日本、新加坡、美国、英国等国家智能鼠比赛，并多次蝉联世界冠军。

The micromouse contest is an integration of multiple disciplines and many technologies. It involves machine engineering, electronic engineering, automatic control, artificial intelligence, program design, sensing and testing technology.

The micromouse contest will enhance the participant's technology level and application abilities, providing a platform for technological innovation.

The publication of books on micromouse education will play a significant role in learning micromouse technology for Chinese students. The micromouse robots made by Qicheng are to the world through Luban workshops, benifitting students around the world.

Congratulations on the publication of the micromouse book series. They provide convenience and reference for micromouse fans and students at all levels.

智能鼠是集多学科多技术的融合体，主要涉及机械工程、电子工程、自动控制、人工智能、程序设计、传感与测试技术等学科。

竞赛的开展，必然提升参赛者在相关领域的技术水平和应用能力，为技术创新提供平台。

这类教育资源书籍的出版，对于中国乃至世界学生学习智能鼠技术都有着重要的意义和作用，启诚电脑鼠通过鲁班工坊冲出国门走向世界，让世界各地的学生受益，我为启诚智能鼠感到骄傲和自豪！

由衷地祝贺智能微型运动装置（Micromouse）技术与应用系列丛书出版。这为智能鼠爱好者、不同教育层面的学生学习嵌入式微型机器人（智能鼠）提供了便利和参考。

Mr. Peter Harrison

英国伯明翰城市大学高级研发工程师。

国际智能鼠走迷宫教育教学专家。

多年来从事设计、研发IT集成项目工作，在培养大学生人工智能机器人技术领域、实训教学和实用技能方面成绩卓著，曾多次参加美国、日本、新加坡、英国等国家的智能鼠走迷宫比赛，并多次蝉联世界冠军。

——Mr. António Valente

智能鼠名人榜

扫码观看

十年前，我开始组织智能鼠比赛，因为我认为这个比赛最适合激发学习者对 STEM 领域的热情。各个年龄和各种水平的选手都可以参加这个比赛。但遗憾的是，有关这个领域鲜有详细的介绍。因此，有关智能鼠的这套书十分有益。作为一个智能鼠爱好者和葡萄牙智能鼠赛事的组织者，我向大家推荐这套书。我认为这套书将会促进智能鼠的发展。

Mr. António Valente

葡萄牙 Trás-os-Montes and Alto Douro 大学科学技术学院教授，高级研究员。

葡萄牙国际智能鼠大赛组委会主席。

研究方向：MEMS 传感器、微控制器和嵌入式系统，重点是农业应用。完成了多项葡萄牙国家级和国际级资助的纵向和横向科研项目和课题，包括：安全、Eno-分析、RobTech、IPAVPSI、Focus 等。

  "智能鼠",英文名为 Micromouse,是使用嵌入式微控制器、传感器和机电运动部件构成的一种智能微型运动装置(嵌入式微型机器人),智能鼠可以在不同"迷宫"中自动记忆和选择路径,采用相应的算法,快速到达所设定的目的地。智能鼠走迷宫竞赛结合了机电一体化、控制论、光学、程序设计和人工智能等多方面的科技知识。

  四十多年来,电气电子工程师学会(IEEE)每年举办一次国际性的智能鼠走迷宫竞赛,自举办以来各国踊跃参加,尤其是美国和欧洲国家的高校学生,为此有些大学还特别开设了"智能鼠原理与制作"的选修课程。中国从2007年开始在上海长三角地区举行小规模尝试性比赛。2009年,天津启诚伟业科技有限公司将这项国际赛事引进天津,以工程实践创新项目(EPIP)教学模式,对智能鼠走迷宫竞赛进行本土化创新改革,对于后期智能鼠竞赛的开展和走进课堂、融入教学起到关键性的作用。经过多年的蜕变与优化,"智能鼠"已经成为集"职业性、综合性、先进性、趣味性"于一体的创新实践教育平台,在推动课程改革、提高教学质量、培养学习者的工程实践创新能力等方面发挥了重要的作用。

  为了将智能鼠的成果进一步推广应用,我们编写了适用于基础教育和职业教育的《智能鼠原理与制作》(初级篇)教材。本书以天津启诚伟业科技有限公司提供的 TQD-Micromouse-JQ 智能鼠为载体,由浅及深、由易到难地进行实践教学。

  本书遵循递进原则,从"玩转"到"掌握",再到"精通",丰富学习者的工程实践知识和技术应用经验,拓展学习者的专业视野,内化形成良好的职业素养,提升学习者的实践创新能力。本书所选案例均来自真实的工程项目,编者均来自国内长期从事智能鼠研究与开发、国际智能鼠走迷宫竞赛获奖的院校和企业。

  本书在重要的知识点、能力点和素养点上,配有丰富视频、图片、文本等资源,学习者可以通过扫描书中二维码获取相关信息。本书编著者长期的国际化教学活动积淀,使得本书成为推进国际化人才培养实践的教学载体,智能微型运动装置(Micromouse)技术与应用系列丛书是天津市"一带一

路"联合实验室（研究中心）——天津中德柬埔寨智能运动装置与互联通信技术推广中心研究成果，同时也是工程实践创新项目（EPIP）教学模式规划教材。本书适合作为基础教育学校开展职业启蒙、科技活动和特色教育的教材，也可作为职业教育的教材，还可作为相关工程技术人员培训用书及智能鼠爱好者参考用书。

在本书附录中提供了"智能鼠原理与制作"（适用于中等职业学校）国际实训课程标准，60学时教学内容，课程设计与实施采用理实一体的方式，以动脑动手相结合，整体贯穿技术技能训练，为职业道德素养内化形成服务。本课程内容与多个国家的"鲁班工坊"建设项目高度融合，服务"一带一路"倡议，推广中国职业教育标准，为"一带一路"沿线国家提供丰富实践教学资源，服务各地技术技能人才培养。

本书由天津大学教授王超，南开大学副教授高艺，启诚智能鼠创始人、天津启诚伟业科技有限公司总经理宋立红编著。英文部分由天津市东丽区职业教育中心学校一级教师王晓芹，天津启诚伟业科技有限公司总经理助理严靖怡，天津机电职业技术学院讲师周繁雨编译。天津市东丽区职业教育中心学校高级教师李鑫，天津市东丽区职业教育中心学校高级教师刘学文，天津市东丽区职业教育中心学校高级教师董庆林，天津市东丽区职业教育中心学校一级教师李军，天津市东丽区职业教育中心学校一级教师张宏成，南开大学电子信息与光学工程学院实验师李晓晨参与了本书编写和翻译工作。国际专家美国麻省理工学院教授David Otten和英国伯明翰城市大学教授Peter Harrison、葡萄牙Trás-os-蒙特斯与奥拓杜罗大学教授António Valente参与本书英文内容的译审，并专为本书写了贺信。本书在编写过程中得到了天津大学、南开大学、天津市东丽区职业教育中心学校、天津第一轻工业学校、天津经济贸易学校、美国麻省理工学院、英国伯明翰城市大学和葡萄牙Trás-os-蒙特斯与奥拓杜罗大学等相关院校教授、专家的大力支持。天津启诚伟业科技有限公司陈立考、邱建国、宋姗为本书出版提供了企业实际工程案例、二维码视频、PPT等课程资源。衷心感谢天津市教育委员会、中国铁道出版社有限公司、天津启诚伟业科技有限公司对本教学资源开发提供的指导与帮助。本书由天津市东丽区职业教育中心学校赞助编译，中国铁道出版社有限公司支持出版，并通过鲁班工坊在"一带一路"沿线国家使用。

限于编著者的经验、水平以及时间，书中难免存在不妥和缺漏，敬请专家、广大读者批评指正。

编著者

2020年8月

目录

第一篇　基础知识篇...**001**

项目一　智能鼠的发展历程.. 003

任务一　智能鼠的起源.. 003

任务二　智能鼠的竞赛与调试环境............................... 009

项目二　智能鼠的硬件结构.. 012

任务一　智能鼠的组成.. 012

任务二　智能鼠大脑的结构... 014

项目三　智能鼠的开发环境.. 016

任务一　Arduino软件开发环境...................................... 016

任务二　智能鼠的逻辑思维方式................................... 019

项目四　智能鼠的基础功能调试... 024

任务一　智能鼠的眼睛看世界.. 024

任务二　智能鼠的腿动起来... 028

第二篇　综合实践篇...**031**

项目一　智能鼠的交互控制.. 033

任务一　智能鼠人机交互系统.. 033

任务二　智能鼠眼睛和腿的协同工作........................ 037

项目二　智能鼠的姿态控制.. 039

任务一　智能鼠跑起来.. 039

任务二　智能鼠学会转弯... 046

第三篇　竞赛创新篇...**051**

项目一　智能鼠常用算法解析... 053

任务一　左手法则.. 053

任务二　右手法则.. 054

项目二　智能鼠高级功能.. 056

任务一　多传感器协同工作原理与实现 056

任务二　多线程工作原理与实现 061

任务三　智能鼠躲避障碍灵活运行 063

第四篇　拓展应用篇 **071**

项目一　TQD-IOT工程创新课程平台结构组成 073

任务一　工程创新课程平台和智能鼠的关系 073

任务二　工程创新课程平台的硬件组成 075

项目二　智能鼠技术IOT扩展应用 079

任务一　智能鼠控制器对灯光系统的控制 079

任务二　智能鼠控制器对安防系统的控制 086

任务三　智能鼠控制器对显示系统的控制 098

附录 **109**

附录A　风靡全球的国际智能鼠走迷宫竞赛 111

附录B　入门级智能鼠竞赛案例分析 120

附录C　TQD-Micromouse-JQ器件清单 125

附录D　TQD-IOT工程创新课程平台器件清单 125

附录E　教学内容和学时安排 126

附录F　电路图形符号对照表 126

附录G　专业词汇中英文对照表 127

附录H　"智能鼠原理与制作"国际实训课程标准 129

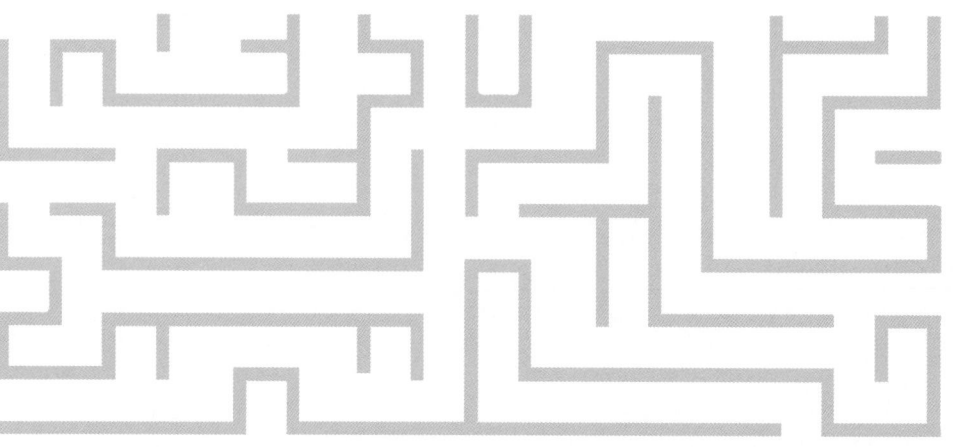

第一篇　基础知识篇

  智能鼠走迷宫竞赛在国际上已经有40多年的历史，竞赛要求智能鼠从起点出发，在不受人为操纵影响的条件下在未知的迷宫中，自主搜索迷宫找到终点，并挑选出最短的一条路径进行冲刺。竞赛根据搜索迷宫的时间和冲刺到终点所用时间分出名次。竞赛迷宫遵照电气电子工程师学会（IEEE）的国际标准。在本篇中，将分别从国际IEEE标准迷宫场地、智能鼠的硬件系统和软件开发环境等方面系统介绍智能鼠技术，并对智能鼠的基本原理和实际操作方法进行具体说明。

项目一

智能鼠的发展历程

学习目标

（1）了解智能鼠的发展历程。

（2）理解智能鼠走迷宫竞赛平台——竞赛迷宫场地、全自动计分系统。

任务一　智能鼠的起源

一、智能鼠的起源

1938年，美国密歇根州的数学家香农（Claude Elwood Shannon）完成了《继电器和开关电路的符号分析》的论文。由于布尔代数只有0和1，恰好与二进制对应，香农将它运用于以脉冲方式处理信息的继电器开关，从理论到技术彻底改变了数字电路的设计方向，因此，这篇论文在现代数字计算机史上具有划时代的意义。

1948年，香农又发表了一篇至今还在闪烁光芒的论文——《通信的数学理论》，从而给自己赢得了"信息论之父"的桂冠。

1956年，他参与发起了达特默斯人工智能会议，成为这一新学科的开山鼻祖之一。他不仅率先把人工智能运用于计算机下棋方面，而且还发明了一个能自动穿越迷宫的"智能鼠"，以此证明计算机可以通过学习提高智能。

二、智能鼠的国际发展历程

1972年，《机械设计》杂志发起了一场竞赛。在竞赛中，仅由捕鼠器弹簧驱动的机械鼠，不停地与其他参赛鼠竞赛，以判断哪个机械鼠能够沿着跑道跑出最长的距离。

1977年，IEEE Spectrum 杂志提出智能鼠的观念。智能鼠是一个小型的由微处理器控制的机器人车辆，在复杂迷宫中具有译码和导航的功能和能力。

1979年，电气电子工程师学会（IEEE）通过其Spectrum and Computer杂志

发起了一场智能鼠竞赛，奖励能够在最短时间内自主走出迷宫的智能鼠的设计者1 000美元。

1980年，东京举办了首场全日本Micromouse国际公开赛，之后，又有多个比赛被创办，如：1980年英国智能鼠大赛，1987年新加坡举办了第一届新加坡Micromouse竞赛和2007年中国计算机学会举办的首场IEEE国际标准Micromouse走迷宫竞赛等，如图1-1-1所示。

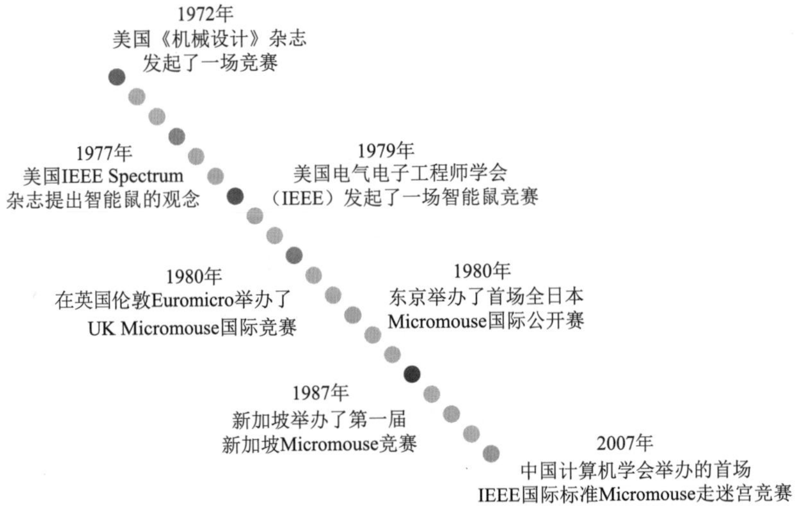

1972年
美国《机械设计》杂志
发起了一场竞赛

1977年
美国IEEE Spectrum
杂志提出智能鼠的观念

1979年
美国电气电子工程师学会
（IEEE）发起了一场智能鼠竞赛

1980年
在英国伦敦Euromicro举办了
UK Micromouse国际竞赛

1980年
东京举办了首场全日本
Micromouse国际公开赛

1987年
新加坡举办了第一届
新加坡Micromouse竞赛

2007年
中国计算机学会举办的首场
IEEE国际标准Micromouse走迷宫竞赛

图1-1-1　智能鼠国际发展

从最初1972年的机械电子鼠发展到现在的智能鼠，经过了40多年的蜕变，参加竞赛的选手从开始仅限于哈佛大学、麻省理工学院等世界知名学府的研究生，发展到从研究型大学到应用技术大学再到职业院校的学生，甚至是中小学生。多教育层面都采纳智能鼠作为教学载体，培养学生们的工程素养以及科技创新意识、动手设计能力。

各类智能鼠竞赛也如雨后春笋般蓬勃发展。目前智能鼠竞赛已经成为应用于不同教育阶段的国际创新型学生竞赛。

三、智能鼠的中国发展历程

从2007年至今，智能鼠在中国经历了十余年的发展历程，如图1-1-2所示。2007年天津启诚伟业科技有限公司将这项国际赛事引进天津，以中国先进的教育模式"工程实践创新项目"为核心理念，对智能鼠走迷宫竞赛进行本土化创新改革，助力智能鼠竞赛在中国的蓬勃开展，对于智能鼠技术走进课堂融入教学起到关键性的引领作用。

视 频
中国智能鼠发展

首场比赛	高等教育	职业教育	普通教育	国际竞赛
2007年 中国首场智能 鼠走迷宫竞赛	2009年至今 大学生学科竞赛 Micromouse大赛	2010年至今 职业院校学生技能竞赛 Micromouse大赛	2016年至今 普职融通国际挑战赛 Micromouse大赛	2016年至今 中国IEEE智能鼠 国际邀请赛

图1-1-2　智能鼠在中国的发展

竞赛对于满足产业优化升级，开阔国际视野，掌握实践与创新经验，培养高技术、高技能人才，起到了引领推动作用（见图1-1-3）。智能鼠在中国从大学生竞赛到职业院校大赛，再到普职融通国际挑战赛，积累了丰富的竞赛经验和优秀的技术积淀。

图1-1-3　竞赛纪实照片

十余年来，中国的智能鼠竞赛不断创新国际发展新思路，从最初的"简单模仿"学习，发展到目前的"互学互鉴"，逐步搭建起国际交流合作的新平台，先后经历了学习借鉴、蜕变升华和引领辐射三个阶段。

首先是学习借鉴：2015年天津大学生代表队征战美国第30届APEC世界Micromouse竞赛（见图1-1-4），取得了世界排名第六的好成绩。2017年至2018年，天津启诚伟业科技有限公司全额资助了在天津大学生智能鼠竞赛上获得企业命题赛冠军队，到日本东京参加第38届和第39届全日本Micromouse国际公开赛（见图1-1-5），促进学习借鉴国际智能鼠先进技术，结识众多智能鼠业界专家教授，对中国智能鼠技术的发展与提升起到推动的作用。

接着是蜕变升华：智能鼠大赛在中国进行本土化创新改革，设计了一系列从易到难的启诚智能鼠教学平台，满足"中、高、本、硕"不同学习阶段学生学习应用。从2016年开始先后邀请美国麻省理工学院的David Otten教授、中国台湾龙华科技大学苏景晖教授、新加坡义安理工学院黄明吉教授、英国伯明翰城市大学Peter Harrison教授、日本智能鼠国际公开赛组委会秘书长中川友纪子先生等智能鼠专家和来自泰国、印度、印尼、巴基斯坦、柬埔寨等国际"鲁班

视频 ●

学习借鉴
（美国
APEC）

视频 ●

蜕变升华
（第一届
IEEE）

工坊"师生，以及来自天津、北京、河南、河北等国内省市精英级代表队，先后加盟中国IEEE智能鼠走迷宫国际邀请赛（见图1-1-6、见图1-1-7）。国际选手通过参加中国比赛，对中国竞赛标准、竞赛规则、竞赛模式和竞赛理念有了更深层次的了解和认同，从而切实推动了国际化的交流与合作，达到"互学互鉴"的目的。

图1-1-4　中国天津代表队远赴美国参加国际大赛

图1-1-5　中国天津代表队远赴日本参加国际大赛

图1-1-6　第三届IEEE智能鼠走迷宫国际邀请赛

图1-1-7　"启诚杯"第四届IEEE智能鼠走迷宫国际邀请赛

最后是引领辐射：教育对外开放是我国改革开放事业的重要组成部分，随着"一带一路"倡议的推进，2016年以来在中国教育部指导下，先后启动了海外鲁班工坊国际项目，智能鼠作为中国优秀的教育装备，伴随着鲁班工坊走出国门与世界分享。从2016年至今，启诚智能鼠来到泰国、印度、印尼、巴基斯坦、柬埔寨、尼日利亚、埃及等国家，免费开展智能鼠竞赛的推广和课程培训，受到了沿线国家师生的一致青睐（见图1-1-8~图1-1-13）。智能鼠成为连接世界的纽带与桥梁！

视　频

引领辐射
（印度鲁
班工坊）

图1-1-8　印度鲁班工坊开展智能鼠培训课程

图1-1-9　2016年泰国鲁班工坊开展智能鼠培训课程

图1-1-10　2017年印尼鲁班工坊开展智能鼠培训课程

图1-1-11　2018年巴基斯坦鲁班工坊开展智能鼠培训课程

图1-1-12　2018年柬埔寨鲁班工坊开展智能鼠培训课程

图1-1-13　2020年埃及鲁班工坊开展智能鼠培训课程

任务二 智能鼠的竞赛与调试环境

一、竞赛迷宫场地

目前，国际和国内比赛都使用同样规格的比赛场地，即一个由8×8个格子组成的方形迷宫。迷宫的"墙壁"是可以插拔的，这样就可以形成各种各样的迷宫。

如图1-1-14所示为TQD-Micromouse Maze 8×8比赛场地。迷宫底板的尺寸为2.96 m×2.96 m，上面共有8×8个标准迷宫单元格。图1-1-15所示为古典智能鼠迷宫挡板和立柱。

图1-1-14 TQD-Micromouse Maze 8×8
迷宫场地

图1-1-15 古典智能鼠迷宫挡板和立柱

TQD-Micromouse Maze 8×8迷宫场地规范如下：

（1）迷宫由8×8个、18 cm×18 cm大小的正方形单元所组成。

（2）迷宫的挡板高5 cm，厚1.2 cm，因此两个挡板所构成的通道的实际距离为16.8 cm，挡板将整个迷宫封闭。

（3）迷宫挡板的侧面为白色，顶部为红色。迷宫的地面为木质，颜色为哑光黑。挡板侧面和顶部的涂料能够反射红外线，地板能够吸收红外线。

（4）迷宫的起始单元可设在迷宫四个角之中的任何一个。起始单元必须三面有挡板，只留一个出口。迷宫的终点设在迷宫中央，由四个正方形单元构成。

（5）在每个单元的四角可以插上一个小立柱，其截面为正方形。如图1-1-14所示。立柱长1.2 cm、宽1.2 cm、高5 cm。小立柱所处的位置称为"格点"。除了终点区域的格点外，每个格点至少要与一面挡板相接触。

（6）迷宫制作的尺寸精度误差应不大于5%，或小于2 cm。迷宫地板的接缝不能大于0.5 mm，接合点的坡度变化不超过4°。挡板和立柱之间的空隙不大于1 mm。

（7）起点和终点设计遵照IEEE智能鼠竞赛标准，即智能鼠按照顺时针方向开始运行。

二、专用测试场地

专用测试场地上绘有13个标记位置，并且使用不同的颜色进行区分（见图1-1-16），用于调试红外传感器和优化转弯控制参数。接下来就带领读者认识一下它：

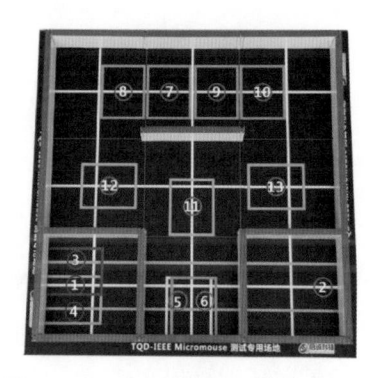

图1-1-16　TQD-IEEE Micromouse
专用测试场地

（1）①至②，灰色通道，用来检测智能鼠在无红外校准的情况下直行的偏移量。

（2）③深红色矩形，④橙色矩形；③至②、④至②均是用来检验有红外校准时的智能鼠直行情况。

（3）⑤黄色矩形用来调节智能鼠左前红外强度，⑥绿色矩形用来调节智能鼠右前红外强度；校正车姿。

（4）⑦、⑧绿色矩形用来调节智能鼠右后红外强度，⑨、⑩绿色矩形用来调节智能鼠左后红外强度，检测路口。

（5）⑪、⑫、⑬三个蓝色矩形用来调试智能鼠转弯90°。

三、全自动计分系统

为了精确计量智能鼠完成迷宫的时间，需要全自动地计算智能鼠通过起点和终点的时间。图1-1-17所示为由天津启诚伟业科技有限公司设计生产的用于智能鼠走迷宫竞赛的电子自动计分系统。

TQD-Micromouse Timer V2.0系统包含起点对射模块、终点对射模块、智能鼠计分系统模块、计分软件等。

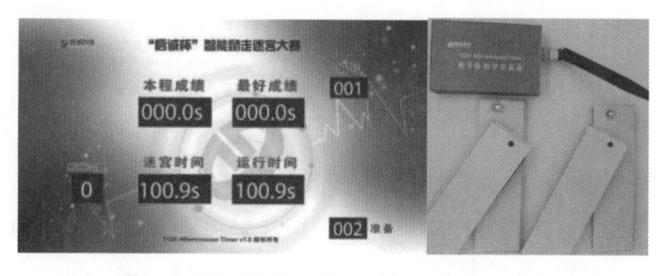

图1-1-17　TQD-Micromouse Timer V2.0

起点对射模块和终点对射模块采用迷你USB充电方式，通过内置的一组激光对射传感器检测智能鼠经过。智能鼠计分系统模块用于接收起点对射模块和终点对射模块通过ZigBee发过来的数据，经过计算机中的计分软件处理，以一

种直观的方式展现智能鼠在迷宫中的运行情况。计分软件也可以单独使用，可通过鼠标输入起点事件和终点事件。计分系统整体的计时精度可达0.001 s。

起点对射模块和终点对射模块分别安装在起点迷宫格和终点迷宫格中，如图1-1-18、图1-1-19所示。当智能鼠经过时，激光被阻断，从而产生起点或终点信号。

图1-1-18 迷宫起点

图1-1-19 迷宫终点

视 频

计分系统工作原理

思考与总结

（1）IEEE国际标准智能鼠场地由哪些部分组成？

（2）请归纳智能鼠竞赛的特点。

（3）全自动计分系统大大提高了竞赛成绩计算的准确性，请简要说明其工作原理。

项目二

智能鼠的硬件结构

学习目标

（1）认识智能鼠的基本硬件结构。

（2）学习智能鼠硬件之间是如何工作的。

本书以TQD-Micromouse-JQ智能鼠为教学载体，如图1–2–1所示。TQD-Micromouse-JQ智能鼠以国际创客届主流的Arduino为核心处理器，提供图形化软件编程环境，提升初学者的学习兴趣；它适用于智能鼠走迷宫竞赛入门级学生实训教学，是培养学生工程素养和科技创新能力的首选平台。

图1–2–1　TQD-Micromouse-JQ智能鼠

任务一　智能鼠的组成

一、TQD-Micromouse-JQ智能鼠的电路组成

TQD-Micromouse-JQ智能鼠主要有以下结构特点：

（1）主控芯片采用高性能低功耗的AVR ATmega328P；

（2）内置HC06蓝牙模块，实时高速、准确传输数据，有效距离可达10m以上；

（3）App在线手机调试，灵活方便，实现智能鼠与移动蓝牙无线通信功能；

（4）五组高精度红外数字传感器，全方位检测迷宫挡板信息；

（5）采用钢齿轮精密N20减速电动机，工作电压为3~6 V，抗干扰能力强；

（6）TQD-Micromouse-JQ智能鼠车体外壳采用 3D打印 DIY设计，自由拼接，创意无限；

（7）基于国际开源的 Arduino易学、易懂、易练的图形化编程，使学习生动有趣；

（8）TQD专用DEMO程序包，包含现场调试、优化算法等程序，让零基础学习者快速上手，迅速提升能力。

TQD-Micromouse-JQ智能鼠的电路组成框图如图1-2-2所示，主要包括核心控制电路、电源电路、红外检测电路、蓝牙电路、电动机驱动电路和数字/模拟扩展接口这六个部分，其中最主要的部分为核心控制电路，其他组成部分都是依靠它来工作的。电源电路驱动核心控制电路，然后由其与其他几个部分进行交互，实现智能鼠的功能。

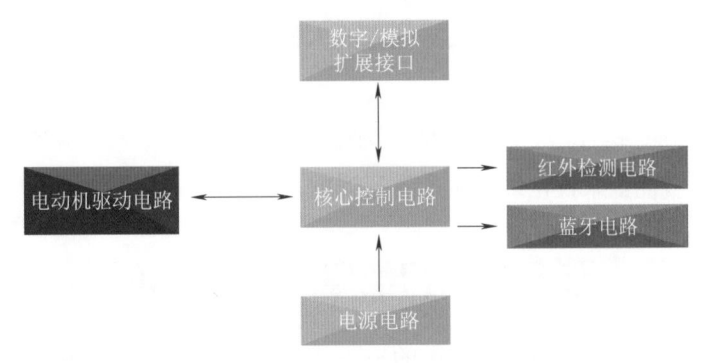

图1-2-2 TQD-Micromouse-JQ智能鼠的电路组成框图

二、TQD-Micromouse-JQ智能鼠的框架结构

读者在接触本书之前，对于智能鼠这个词或许并没有什么概念，但是如果说到机器人，大家一定不陌生。那么机器人是如何定义的呢？可以说，机器人是一种自动化的机器，能够实现某种或某些特殊任务，这种机器具有一定的智能，比如具有一定的感知能力、规划能力、动作能力、协同能力等。就这个意义来说，智能鼠的本质就是一种微型智能移动机器人。

虽然机器人具有的智能和实现的任务各不相同，但它们的结构大致都可以划分为几部分，即传感器部分、主控芯片部分、执行器部分。

TQD-Micromouse-JQ智能鼠的元件布局图如图1-2-3所示。

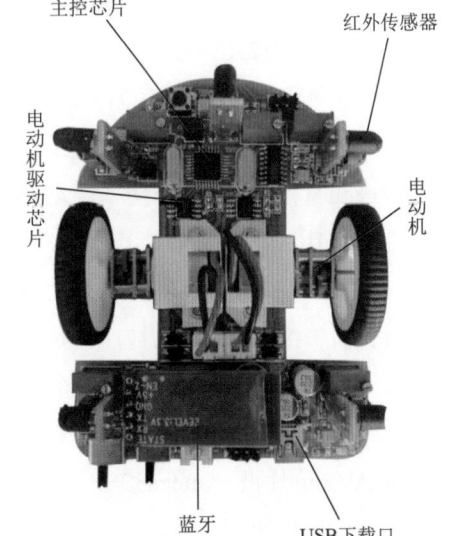

图1-2-3 TQD-Micromouse-JQ智能鼠的元件布局图

视频

智能鼠组成

1. 传感器部分

智能鼠具有一定的智能，为了能够依据外部环境做出反应，需要获取外界信息，包括环境中的声、光、电、磁、温度和湿度、障碍物信息等。用来获取外部信息的传感器就像是人的眼睛、耳朵等感觉器官。

2. 主控芯片部分

主控芯片部分是智能鼠的核心元件，它接收传感器部分传递过来的信号，并根据事前写入的决策系统（软件程序），决定对外部信号的反应，并将控制信号发给执行器部分。它的作用好比人的大脑。

3. 执行器部分

智能鼠通过执行器完成不同的行为或动作，如点亮发光二极管、发出声音等，但对于智能鼠来说，最基本的执行器就是轮胎。它的作用好比人的四肢。

任务二 智能鼠大脑的结构

AVR单片机是Atmel公司推出的较为新颖的单片机，其显著的特点为高性能、高速度、低功耗，内置Flash的RISC（reduced instruction set computer，精简指令集计算机），该单片机具有如下特性：

（1）AVR单片机指令以字为单位，且大部分指令都为单周期指令。在执行当前指令的同时完成下一条指令的读取。通常时钟频率为4～8 MHz，故最短指令执行时间为250～125 ns。

（2）AVR单片机的I/O口资源灵活、功能强大，所有的I/O线全部带可设置的上拉电阻，可单独设定为输入/输出，可设定（初始）高阻输入，具有驱动能力强的特性（可省去功率驱动器件）。具备多种独立的时钟分频器，分别供UART、IIC、SPI使用。其中与8到16位定时器配合的具有多达10位的预分频器，可通过软件设定分频系数提供多种档次的定时时间。

目前，AVR单片机已被广泛应用于计算机外围设备、工业实时控制、仪器仪表、通信设备和家用电器等各个领域。

智能鼠核心板是整个智能鼠控制的大脑中枢，TQD-Micromouse-JQ智能鼠采用的AVR单片机ATmega328P，其芯片引脚图如图1-2-4所示。

1. 核心控制器主要特点

处理器：Atmel Atmega328P。

Digital I/O （数字输入/输出端口）：D0～D13。

Analog I/O（模拟输入/输出端口）：A0~A5。

输入电压：USB接口供电或者5~12 V外部电源供电。

输出电压：支持DC 3.3 V/5 V输出。

其中，Digital I/O中的D3、D5、D6、D9、D10、D11端口可以兼作PWM输出接口。

2. 支持手机在线调试

控制板集成蓝牙模块HC-06（见图1-2-5），通信距离最远可达10 m。

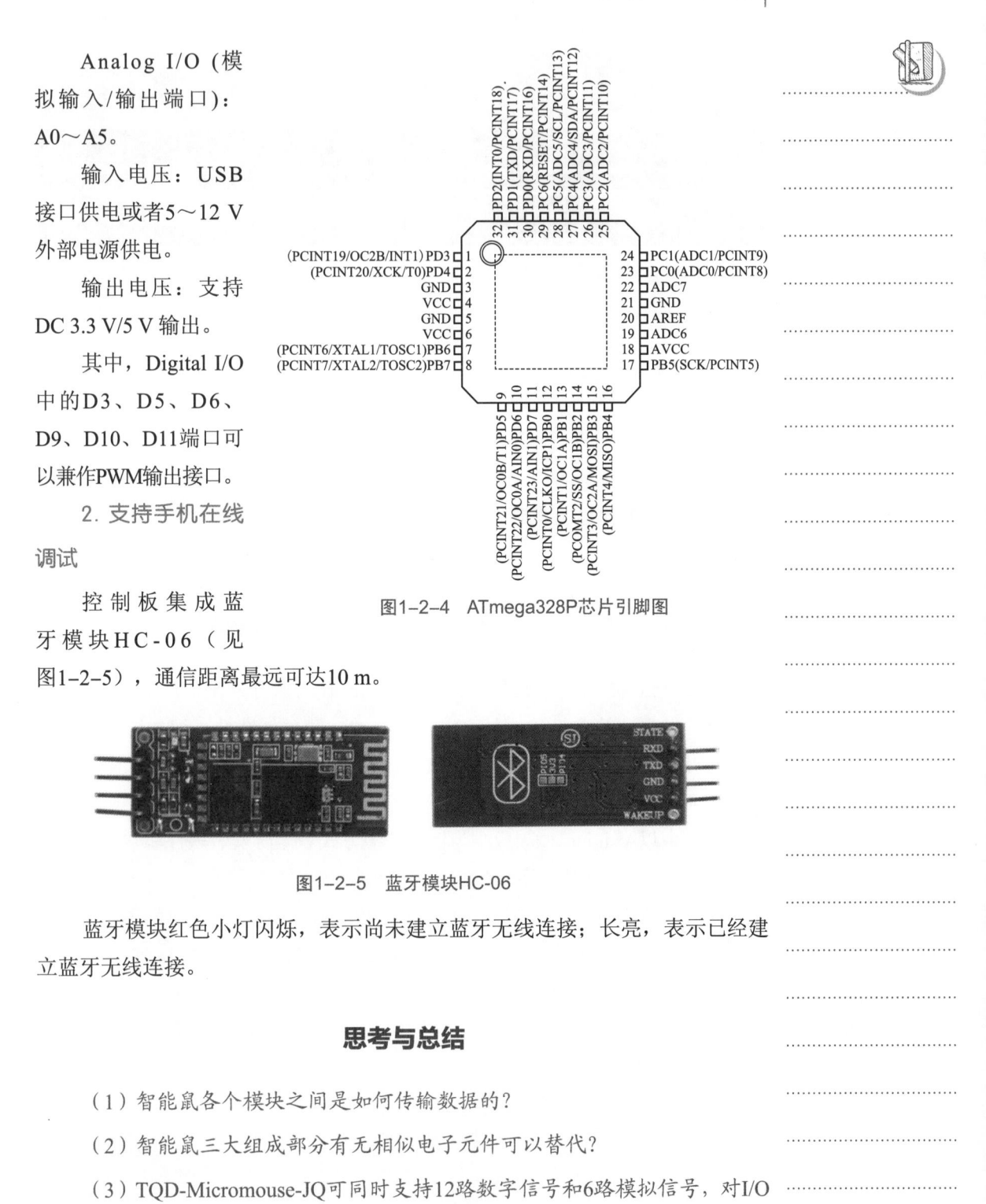

图1-2-4　ATmega328P芯片引脚图

图1-2-5　蓝牙模块HC-06

蓝牙模块红色小灯闪烁，表示尚未建立蓝牙无线连接；长亮，表示已经建立蓝牙无线连接。

思考与总结

（1）智能鼠各个模块之间是如何传输数据的？

（2）智能鼠三大组成部分有无相似电子元件可以替代？

（3）TQD-Micromouse-JQ可同时支持12路数字信号和6路模拟信号，对I/O引脚进行简单设置就可以实现信号的读取和输出。

项目三

智能鼠的开发环境

学习目标

（1）学会使用智能鼠配套软件。

（2）尝试独立编写LED点亮小程序。

硬件设备的运行，离不开软件程序的支撑，编程语言有C、C++、PHP、Python等。这些语言都有一个共同点，即使用英文代码书写程序。作为初学者，使用代码编程存在一定的难度，故本项目介绍一种简单的编程方式——图形化编程。

任务一　Arduino软件开发环境

TQD-Micromouse-JQ智能鼠使用Arduino作为软开发平台，如图1-3-1所示。该开发平台简单易学，具有极强的互动性和可创造性，编程语言极易掌握，同时有着足够的灵活性，不需要很多单片机基础及编程基础，简单学习后，就可以快速进行开发。

本书中以Arduino 1.6.4版本为例，安装方式和平时安装软件非常相似，不需要做任何修改，单击"下一步"按钮即可，直至安装结束。

初次安装后的软件只能输入代码，读者可以输入代码来编写程序。

对于初学者，还提供了另外一种编程方式，即图形化编程，类似于搭建积木。这种方式大大降低了编程的难度，即使是没有编程经验的电子爱好者，包括青少年甚至儿童，都可以尝试着用Arduino控制器，以图形化趣味编程方式，实现自己的设计。

下面为Arduino软件添加图形化编程功能：

（1）选择"文件"→"首选项"命令，如图1-3-2、图1-3-3所示。

● 软件

Arduino开发
环境下载

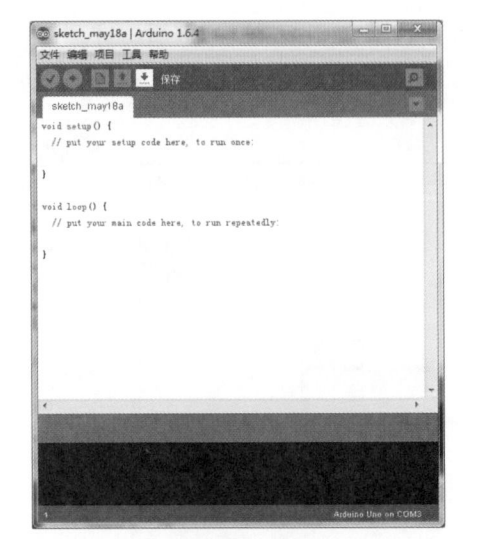

图1-3-1 Arduino软件开发平台

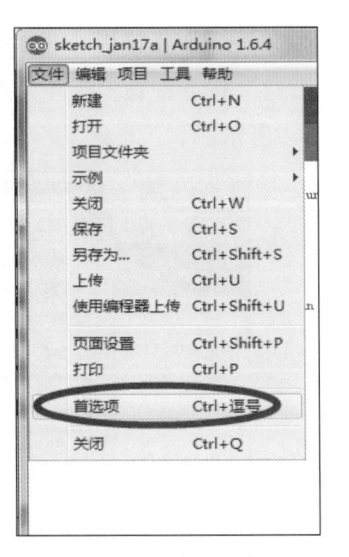

图1-3-2 首选项位置

图1-3-3 项目文件夹位置

（2）复制Jar包。将tools文件夹（在图1-3-4所示的"Ardublock Jar包"文件夹里面），复制到图1-3-5所示的项目文件夹位置。

图1-3-4 Jar包位置

图1-3-5 项目文件夹位置

至此，图形化编程功能已经添加完成，如图1-3-6所示。重启软件，可以在"工具"栏下看到多了一个ArduBlock，单击即可进入图形化编程界面，如图1-3-7所示。

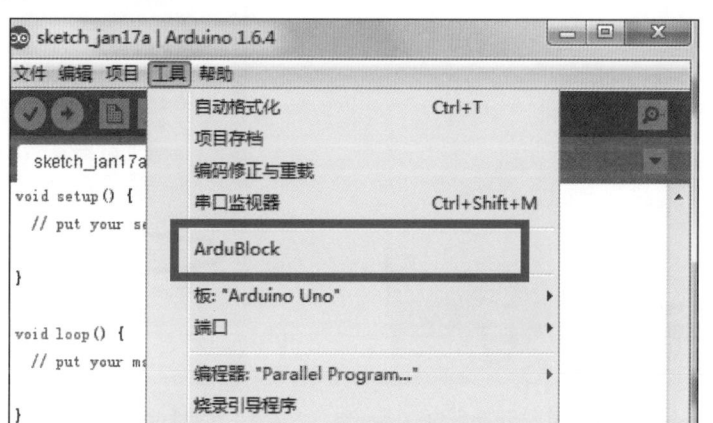

图1-3-6　添加ArduBlock成功

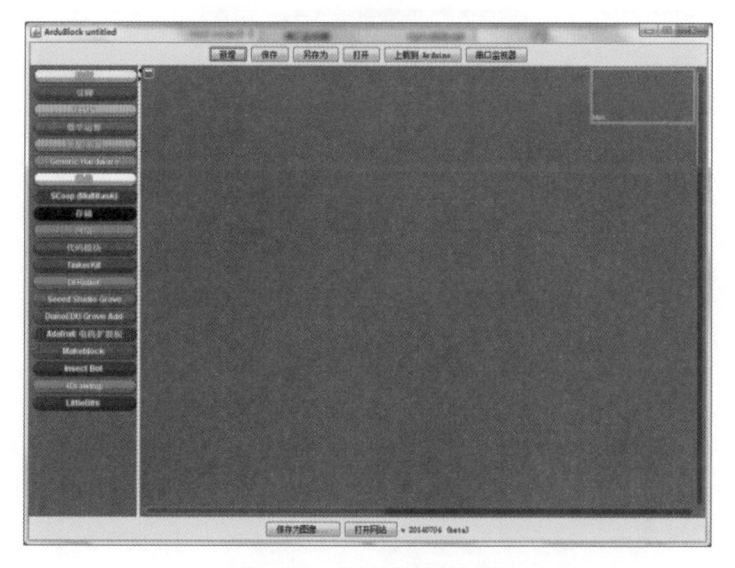

图1-3-7　图形化编程界面

（3）液晶、语音识别和定时中断，Scoop支持库添加。在计算机中找到Arduino的安装位置，并将Arduino/libraries下的LiquidCrystal文件夹完全删除。然后将LiquidCrystal、voiceRecognition和MsTimer2三个文件夹复制到原位置（切勿覆盖，否则需要重新安装Arduino），如图1-3-8所示。

图1-3-8　LCD液晶、语音识别和定时中断支持库

至此，图形化编程功能已经添加完成。重启软件后，就可以开始调用了。

任务二　智能鼠的逻辑思维方式

在Arduino中编写第一个程序——点亮单色LED。

一、图形化编程界面

打开图形化编程界面，在左侧会看到一列不同颜色的栏目，后续实验中会用到的也是最常用的前八个栏目如图1-3-9所示，单击按钮即可查看栏目中的详细内容。

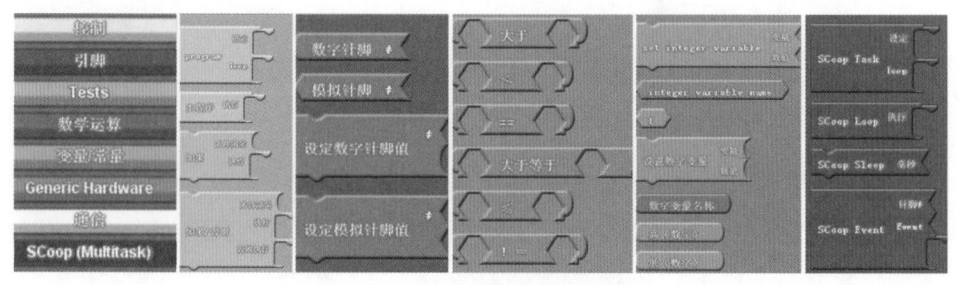

图1-3-9　最常用的前八个栏目

Ardublock图形化编程，针对不同的编程语句设计了不同颜色、不同形状的图形，只有符合编程语法的图形才能够连接到一起。下面针对这八个栏目进行简单介绍。

1. "控制"栏目

"控制"栏目包含主程序、条件判断、循环、延时等对程序运行起到控制

视频 ●

Arduino软件
常用工具栏
介绍

作用的语句。

　　程序的运行首先需要一个主函数，如图1-3-10所示。图1-3-10中的两个主函数模块的不同之处就在于前者多了一个设定，可以设置一些变量的初值等。

　　类似的，这两个条件判断函数模块（见图1-3-11）的区别是当条件不满足时，是否有内容需要被执行。

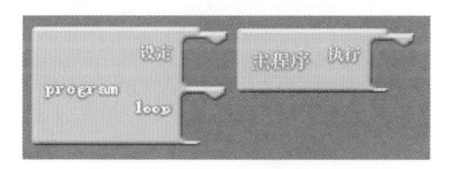
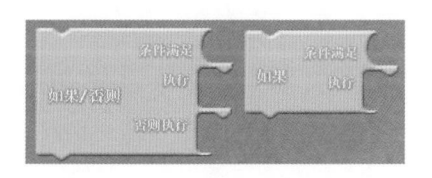

图1-3-10　主函数模块　　　　　　　　图1-3-11　条件判断函数模块

2. "引脚"栏目

　　"引脚"栏目包含传感器和执行器所对应的数字引脚、模拟引脚等。

　　可以看到，根据数据类型的不同，引脚分为两类：数字引脚和模拟引脚（见图1-3-12）。每种模块又分为小模块和大模块，小模块只可以选择引脚号，表示传感器，如按键、温度、光照等。大模块还可以设置高低电平或者模拟量，是用来表示执行器的，如LED、风扇、蜂鸣器等。

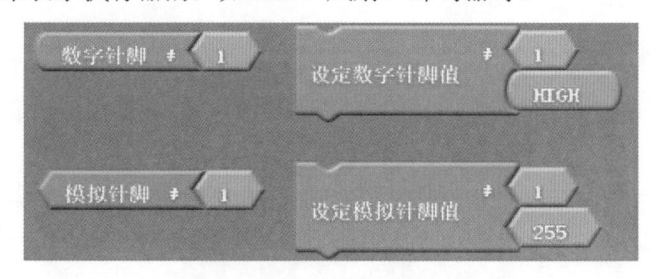

图1-3-12　数字引脚和模拟引脚

3. Tests栏目

　　Tests栏目包含逻辑判断相关的图形。不同的数据类型需要使用不同的图形。

　　整数型——0，1，2，3等整数；

　　字符型——'a'、'b'、'c'和'd'等；

　　数字型——高电平、低电平。

　　不同的数据类型都有自己的形状，这点在使用的时候一定要注意！

4. "数学运算"栏目

　　"数学运算"栏目包含加、减、乘、除等数学运算。当然，可用于数学运算的都必须是整数或整数型变量。

5. "变量/常量"栏目

"变量/常量"栏目包含整数型、字符型、数字型变量的设置，常量的设置等，如图1-3-13所示。从上至下依次为整数型变量、字符型变量、数字型变量。常用的如超声波测得的距离、温度的大小都可以设置为整数型变量；手机App和实验平台进行的无线通信，可以设置为字符型变量；而按键的状态、红外传感器的状态等就需要设置为数字型变量。

6. Generic Hardware栏目

Generic Hardware栏目包含几个通用硬件，如液晶模块、舵机、超声波等。如图1-3-14所示，舵机和超声波使用非常简单，设置正确的角度和引脚号即可。

文本

信号种类——
数字信号、
模拟信号

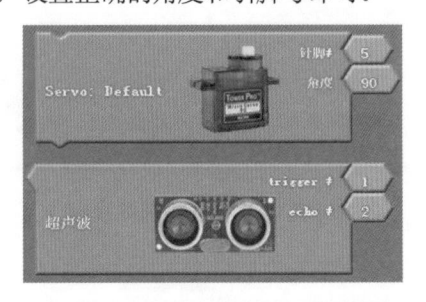

图1-3-13　变量模块　　　　图1-3-14　SG90舵机和超声波模块

液晶模块一定要选用下方的1602（和实验平台一致的型号），选择错误将无法使用。

需要特别注意的是，在数据刷新比较快的时候，还需要添加延时函数以及清屏函数，来辅助读取数据，如图1-3-15所示。

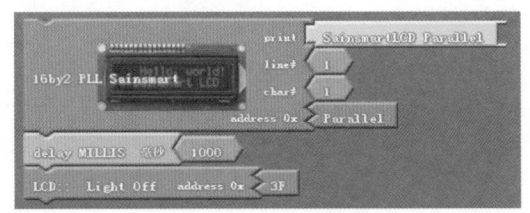

图1-3-15　液晶模块

7. "通信"栏目

"通信"栏目包含读取串口、串行打印等图形，如图1-3-16所示。

图1-3-16　串行打印

读取串口图形用于读取无线通信的数据，如图1-3-17所示。串行打印图形用于回传信息，如图1-3-18所示。

图1-3-17　读取串口数据　　　　　　　　图1-3-18　回传信息

8. Scoop栏目

Scoop栏目主要使用到的模块包括主函数、延时两个模块，如图1-3-19、图1-3-20所示。

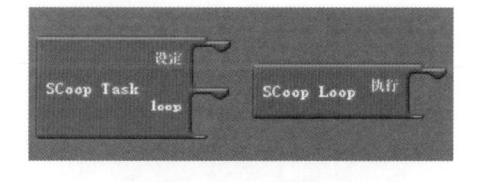

图1-3-19　主函数模块　　　　　　　　图1-3-20　延时模块

了解各个栏目的功能，会对以后的实验有很大的帮助。

下面，尝试编写点亮单色LED模块的程序。实验目的就是给LED输入高电平，使它发光。

二、流程图

LED发光流程图如图1-3-21所示。

三、图形化编程

（1）首先，程序运行是从主函数开始的，所以需要用到"控制"栏中的"主程序"模块，如图1-3-22所示。

图1-3-21　LED发光流程图

● 视频

点亮LED灯

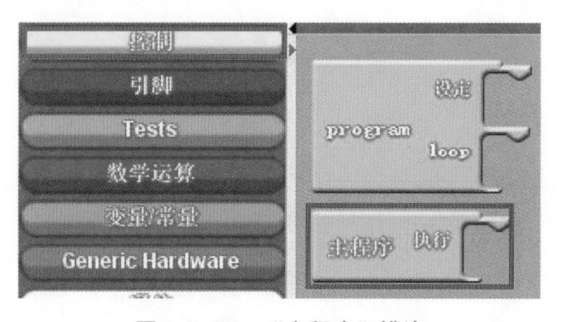

图1-3-22　"主程序"模块

（2）点亮LED是对LED输出高电平信号，需要用到"引脚"栏中的"设定数字针脚值"模块，如图1-3-23所示。

（3）将这两个模块组合起来，将单色LED模块连接到智能鼠核心控制器上，如8号引脚。

对图形化编程模块设置正确的引脚号和高低电平，如图1-3-24所示。

图1-3-23 "设定数字针脚值"模块　　图1-3-24 设置正确的引脚号和高低电平

至此，点亮LED的程序就编写完成了，下面将核心控制器连接到计算机上，并选择对应的串口号，下载程序，观察LED是否点亮，如图1-3-25所示。

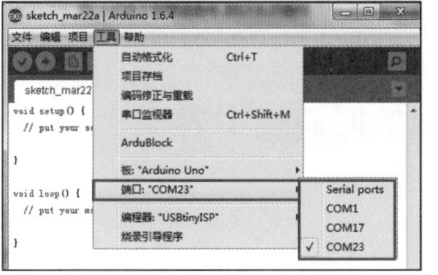
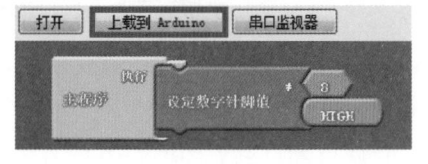

图1-3-25 点亮LED

思考与总结

（1）数字信号和模拟信号的区别是什么？

（2）光照强度和按键信号各属于什么信号？

（3）图形化编程针对代码编程做了封装，不同的图形对应不同的编程语句；形状和颜色各有不同，只有符合编程规则的图形才可以连接在一起，从而大大降低了初学者的编程难度。

项目四

智能鼠的基础功能调试

学习目标

（1）学习智能鼠红外的调试方法。

（2）学习如何驱动智能鼠电动机。

智能鼠的基本功能是从起点运行到终点。在有限宽度、大量转弯的迷宫当中快速而准确行走，离不开高精度的传感器检测以及电动机运行控制。不同的环境，光照强度不同，地面摩擦也有一定的差异，所以必须使用人机交互的手段，调试智能鼠的红外检测精度和电动机转速。

任务一　智能鼠的眼睛看世界

一、机器人传感器的作用

传感器在机器人的控制中起了非常重要的作用，正因为有了传感器，机器人才具备了类似人类的知觉功能和反应能力。感知系统是机器人能够实现自主化的必要部分，而传感器是感知系统中必不可少的部件。

根据传感器的作用，移动机器人中所采用的传感器可以分为内部传感器和外部传感器。内部传感器主要测量机器人内部系统，比如温度、电动机速度、电动机载荷、电池电压等；外部传感器主要测量外界环境，比如距离、声音、光线等。根据传感器的运行方式，移动机器人中所采用的传感器分为被动式传感器和主动式传感器。被动式传感器本身不发出能量，比如CCD、CMOS摄像头传感器；主动式传感器会发出探测信号，比如超声波、红外光、激光，但是此类传感器的反射信号会受到很多因素的影响，从而影响准确信号的获得，同时，信号还很容易受到干扰。

下面介绍如何通过红外传感器实现迷宫挡板的检测。红外传感器是利用三角测量法的原理进行探测的。三角测量法就是把发射器和接收器按照一定角度安装，与被探测点形成一个三角形，由于发射器和接收器的距离已知，发射角

度已知，反射角度也可以被检测到，因此检测点到发射器的距离就可以求出。这个测量法可以测得距离非常近的物体，目前最精确的可以达到1 μm的分辨率，但是不能探测远距离物体。

TQD-Micromouse-JQ智能鼠共安装五组高精度数字红外传感器，如图1-4-1所示。可以通过旋转红外传感器旁边的电位器来改变它的发射强度，当接收头接收被障碍物反射回来的红外线时，从而得出智能鼠四周的挡板信息。控制器根据这些挡板信息，结合逻辑算法，控制智能鼠智能行走。

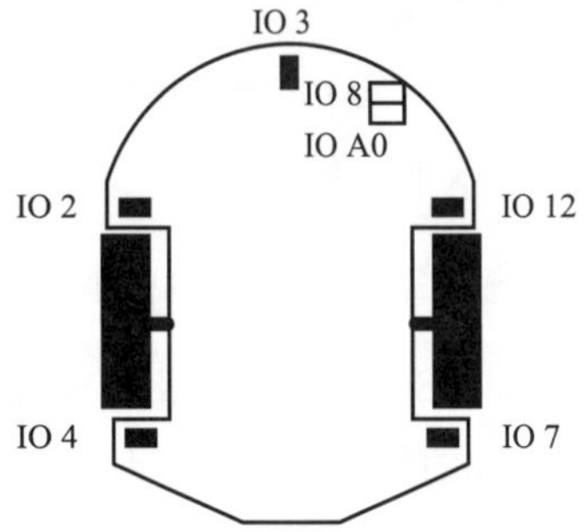

图1-4-1　智能鼠底板示意图

二、智能鼠红外电路组成

智能鼠的红外检测电路，相当于智能鼠的"眼睛"，作为智能鼠系统的输入模块，用于迷宫挡板的检测，分为左前方、正前方、右前方、左后方、右后方五组红外传感器。由上可知，红外发射强度是否准确，是关系到智能鼠能否自动而准确行走的关键。

五组红外传感器的发射端均连接到一个公共发射引脚，即IO_11上，发出特定的PWM波。

五组红外传感器电路原理相同，如图1-4-2所示。

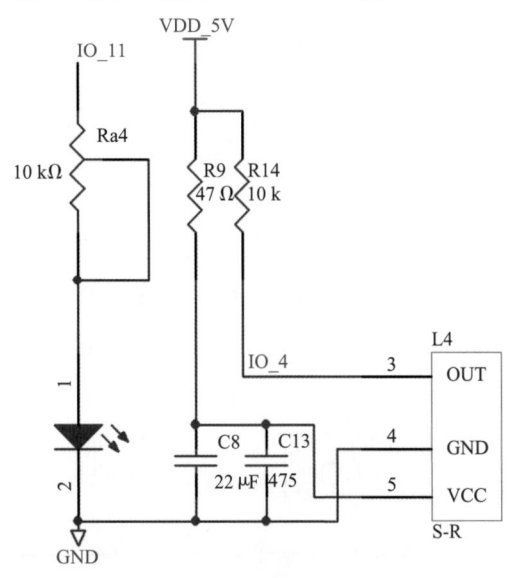

图1-4-2　红外传感器电路原理图[①]

①类似图稿为Protel 99 SE导出的原理图，其图形符号与国家标准符号不一致，二者对照关系参见附录F。

红外传感器的接收头为一体式红外线接收传感器，其型号为IRM8601S，该接收头内部集成自动增益控制电路、带通滤波电路、解码电路及输出驱动电路。该接收头对载波频率为38 kHz的红外线信号最为敏感，当它检测到有效红外线信号时输出低电平，否则输出高电平。Ra4为限流可调电阻，用来调节发射红外线的强度。

三、红外传感器IRM8601S工作原理

文本

传感器背景
知识介绍

IRM8601S是一种工作电压为5 V、接收距离为8 m的红外接收头。外形如图1-4-3所示，其中三个引脚分别为OUTPUT、GND和VCC。

为了便于理解红外接收头的工作原理，首先介绍一下调制的概念。调制是用携带信息的输入信号来控制另一信号的某参数，使之按照输入信号的规律而变化的过程。输入信号称为调制信号，被控制的信号称为载波（或载频）信号，输出信号为调制波，如图1-4-4所示。调制波又根据调制信号所控制的载波信号参数类型分为调幅波、调频波和调相波。

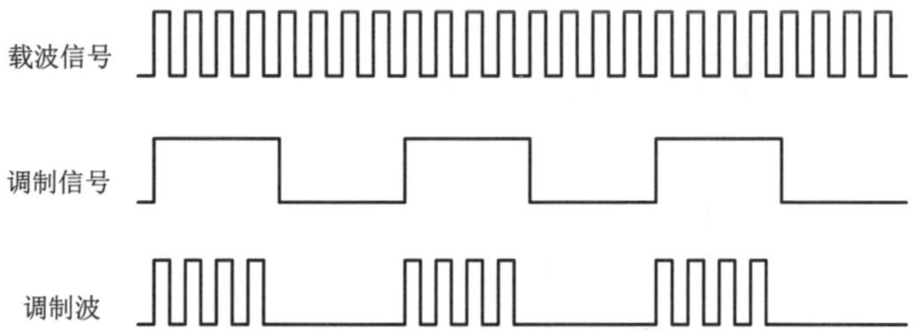

图1-4-3　红外接收头IRM8601S外形

1—OUTPUT；2—GND；3—VCC

载波信号

调制信号

调制波

图1-4-4　调制示意图

IRM8601S传感器内部的带通滤波器的中心频率为38 kHz，所以驱动发射红外线的载波信号为38 kHz时，传感器最灵敏。再根据IRM8601S的数据手册，其调制信号应为周期1 200 μs的方波，驱动红外发射的调制波如图1-4-5所示。

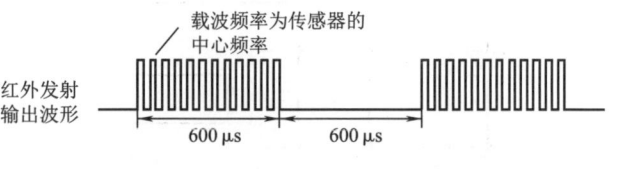

载波频率为传感器的
中心频率

红外发射
输出波形

600 μs　　600 μs

图1-4-5　驱动红外发射的调制波

IRM8601S检测到信号时，输出有效电平（低电平），有效电平维持时间

T_{WL}的范围为400 μs<T_{WL}<800 μs，传感器输出波形如图1-4-6所示。

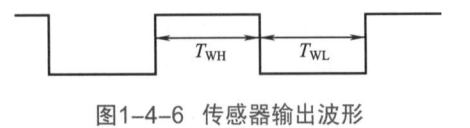

图1-4-6　传感器输出波形

T_{WH}：一个周期内高电平持续时间；

T_{WL}：一个周期内低电平持续时间。

四、红外发射强度调节

想要智能鼠在迷宫当中能够正常顺利地行走，首先需要调节它的前后四个红外传感器。

使用专用测试场地，只需要将智能鼠放置在合适的标识位置，就可以量化调节红外传感器强度，彻底解决调试红外传感器不准确的情况，使红外传感器的调试更简单、更科学。当智能鼠在专用测试场地行走时，会根据红外传感器检测的结果，自动校正车姿和检测路口并转弯，实现智能鼠的智能化。

在开始调试红外传感器发射强度之前，先编写一个小程序，当传感器检测到挡板时，发送数据给手机，从而告诉我们，智能鼠是否"看到"障碍物。

（1）在Arduino软件的应用部分，已经学习过如何发出一定频率的PWM波（使用"有音"和"无音"图形）。

结合红外接收头IRM8601S的特性，发射时长应该保持在600 μs左右，并且一定要关闭发射相同的时间。由于PWM波频率为38 kHz时，红外传感器检测距离最远，先任取一个20～38 kHz之间的频率，例如28 kHz，在后续调节发射强度时，根据实际情况适当修改发射频率，如图1-4-7所示。

（2）下面为四个传感器进行变量命名，如图1-4-8所示。

图1-4-7　红外PWM波驱动

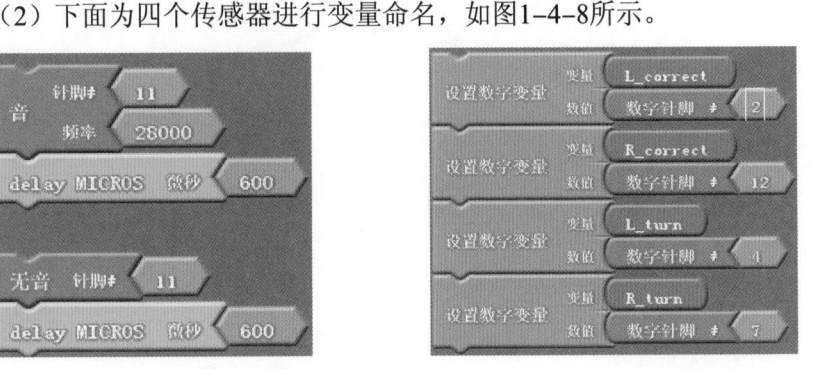

图1-4-8　设置传感器引脚

L_correct：左前方传感器，对应IO 2；

R_correct：右前方传感器，对应IO 12；

L_turn：左后方传感器，对应IO 4；

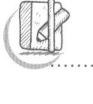

R_turn: 右后方传感器，对应IO 7。

添加四个"如果"图形来判断红外传感器是否检测到挡板，如图1-4-9所示。

图1-4-9　红外测试

任务二　智能鼠的腿动起来

一、电动机背景介绍

● 文本

运动结构背景知识介绍

电机是指依据电磁感应定律实现电能的转换或传递的一种电磁装置，可分为电动机和发电机。电动机是将电能转换为机械能，发电机是将机械能转换为电能。电动机在电路中用字母M（旧标准用D）表示。它的主要作用是产生驱动转矩，作为用电器或各种机械的动力源。电动机按照运转速度可划分为高速电动机、低速电动机、恒速电动机、调速电动机。低速电动机又分为齿轮减速

电动机、电磁减速电动机、力矩电动机和爪极同步电动机等。

TQD-Micromouse-JQ智能鼠采用升级版高品质钢齿轮N20减速电动机（俗称"马达"）作为它的动力源泉，如图1-4-10所示。

二、驱动N20减速电动机

经过之前的学习，对于Arduino软硬件开发都有了一个比较深入的了解。下面可以先经过简单调试使电动机动起来。

N20减速电动机使用简单，两根引脚线的输入电压一高一低时，能够实现电动机的正转和反转，如图1-4-11所示。

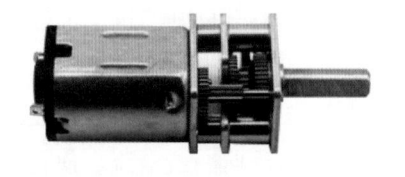

图1-4-10　N20减速电动机

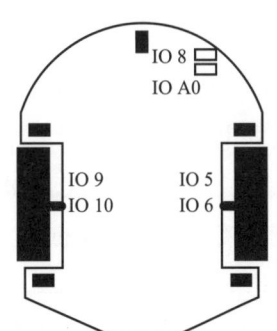

图1-4-11　N20减速电动机引脚号

左侧电动机占用9号和10号引脚，右侧电动机占用5号和6号引脚。其转向真值表见表1-4-1。

表1-4-1　N20减速电动机转向真值表

引脚	9引脚低电平	9引脚高电平	引脚	6引脚低电平	6引脚高电平
10引脚低电平	停	正转	5引脚低电平	停	正转
10引脚高电平	反转	停	5引脚高电平	反转	停

在图1-4-12所示的程序中，向9号引脚输入高电平，向10号引脚输入低电平，下载程序后可以看到，电动机是逆时针旋转的，习惯上称为正转。同样的，可以反过来，向9号引脚输入低电平，而向10号引脚输入高电平，这时电动机会顺时针旋转，也就是反转。不同的电动机模块，正反转的方向可能会不同。

不仅可以控制电动机的正转和反转，也可以控制它转动的速度大小。不过这就要利用PWM技术，输出模拟量进行调节了。

关于智能鼠的两个电动机，现在已经熟悉了它们的运转方式。而在其他的智能鼠中也有采用三轮车式、履带式、四足式等驱动方式。

● 视 频

智能鼠
动起来

三、电动机驱动芯片

L9110S是一种半导体集成产品，极限参数为800 mA/2.5～12 V，宽电源电压范围：2.5～12 V。其引脚图如图1-4-13所示。

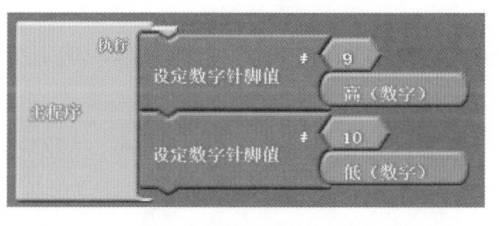

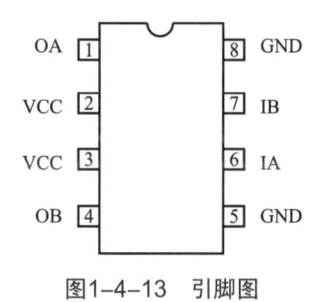

图1-4-12　电动机模块旋转方向测试程序　　　　图1-4-13　引脚图

L9110S是为控制和驱动电动机设计的两通道推挽式功率放大专用集成电路器件，将分立电路集成在单片集成电路之中，使外围器件成本降低，整机可靠性提高。该芯片有两个TTL/CMOS兼容电平的输入，具有良好的抗干扰性；两个输出端能直接驱动电动机的正反向运动，它具有较大的电流驱动能力，每个通道能通过800 mA 的持续电流，峰值电流能力可达1.5 A；同时它具有较低的输出饱和压降；内置的钳位二极管能释放感性负载的反向冲击电流，使它在驱动继电器、直流电动机、步进电动机或开关功率管的使用上安全可靠。L9110S被广泛应用于玩具汽车电动机驱动、脉冲电磁阀门驱动、步进电动机驱动和开关功率管等电路上，如图1-4-14所示。

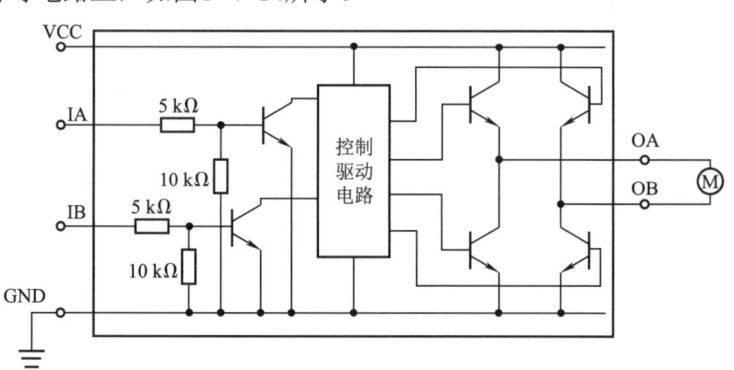

图1-4-14　L9110S电动机驱动电路

思考与总结

（1）红外传感器的作用是什么？如何提高检测精度？

（2）电动机共有哪几种类型？

第二篇　综合实践篇

　　本篇主要针对智能鼠的各个模块进行功能性介绍。主要从智能鼠人机交互系统，了解内置的蓝牙模块和移动端的蓝牙App进行实时的数据传输和信息回传。了解智能鼠传感器和电动机协同工作的原理。智能鼠的姿态控制，驱动电动机运行，实现智能鼠转弯。

项目一

智能鼠的交互控制

学习目标

（1）学习使用蓝牙技术遥控智能鼠。

（2）学习使用传感器控制电动机，实现智能鼠的自动控制。

智能鼠在运行时，应该具备两个控制系统；人机交互控制，主要是智能鼠的启动、停止以及需要紧急应对的情况；自动控制，主要是智能鼠在运行时自动实现系统控制，这样可以提升智能鼠系统的响应速度，节省人力成本。

任务一 智能鼠人机交互系统

一、人机交互的概念

随着人类认知领域的不断扩大和深入，认知层次的不断发展和提高，人类发展空间不断扩大，许多工作具有不可预测性，因此，人机交互系统的开发和应用具有很大的迫切性。嵌入式微型机器人的"人机交互"指的是Human-Robot，即人与机器人之间的交互，通过计算机输入、输出设备，以有效的方式实现人与计算机对话的技术。人机交互技术包括机器通过输出或显示设备给人提供大量有关信息及提示请示等，人通过输入设备给机器输入有关信息、回答问题及提示请示等。人机交互技术是计算机用户进行界面设计的重要内容。它与认知学、人机工程学、心理学等学科领域有密切的联系。

人机交互系统的设计应符合以下原则：可控制原则、易用性原则、直观性原则、简洁性原则和可视性原则。一个好的机器人人机交互系统应使人脑的决策更快地传给机器人，同时也应该更快地把系统信息反馈回来，以便使用者做出决策。

二、TQD-Micromouse-JQ智能鼠人机交互

智能鼠的人机交互系统，是指内置的蓝牙模块和移动端的蓝牙App进行实

软件 ●

蓝牙App
下载

时的数据传输，包括指令发送以及信息回传。

下面通过编程来进行人机交互：当发送指令"a"时，回传信息"Who are you？"；当发送指令"b"时，回传信息"I can see the world！"

1. 蓝牙串口软件

蓝牙是从Bluetooth这个英语单词翻译过来的，它是指一种技术标准，可实现设备之间的短距离数据交换。需要在手机上安装一个蓝牙串口软件，这个软件可以帮助给开发板发送指令。

蓝牙串口软件有很多种，可以选择其中的一种下载安装。安装的方法和安装其他App的方法是一样的。常用的一款蓝牙串口助手主界面如图2-1-1所示。

使用蓝牙串口软件和核心控制器内置的蓝牙模块进行连接，选择正确的设备，并输入配对密码（默认为0000或1234）即可连接成功，如图2-1-2所示。

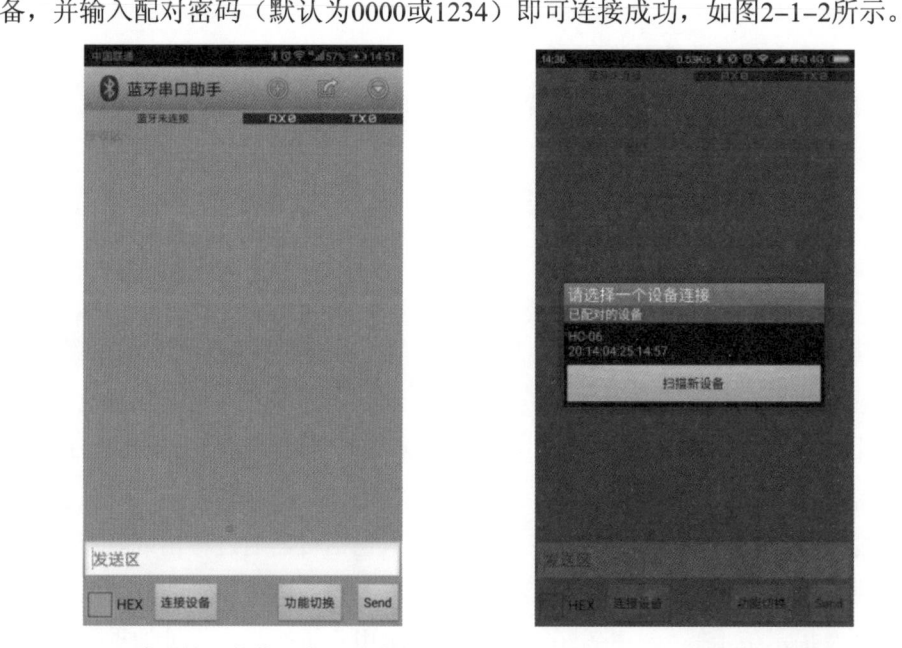

图2-1-1　常用的一款蓝牙串口助手主界面　　　　图2-1-2　连接蓝牙界面

2. 梳理逻辑思路——流程图

现在已经了解了智能鼠的蓝牙人机交互，那么它究竟是如何实现的呢？

首先，需要读取蓝牙数据并接收指令；

其次，要判断指令是什么内容；

最后，根据不同指令，输出不同的信息。流程图如图2-1-3所示。

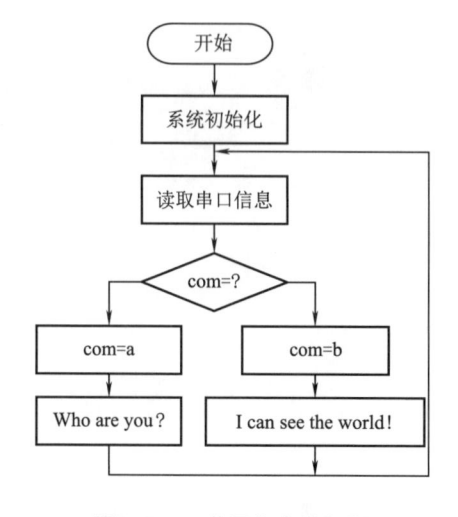

图2-1-3　蓝牙指令流程图

3. 图形化编程

按照之前的分析，需要蓝牙模块读取手机发送的指令，当读到指令"a"时，回传信息"Who are you？"；当读到指令"b"时，回传信息"I can see the world！"。下面就以这个逻辑顺序编写程序。

首先，将"主程序"模块拖动到编程区，在读取串口之前先要判断串口有无数据，如图2-1-4所示。

图2-1-4　判断串口有无数据

其次，设置字符型变量来保存读取到的蓝牙指令。为了便于理解，将变量名重命名为"com"，如图2-1-5所示。

图2-1-5　将变量名重命名为"com"

最后判断指令是"a"还是"b"，如图2-1-6所示，并回传信息，如图2-1-7所示。

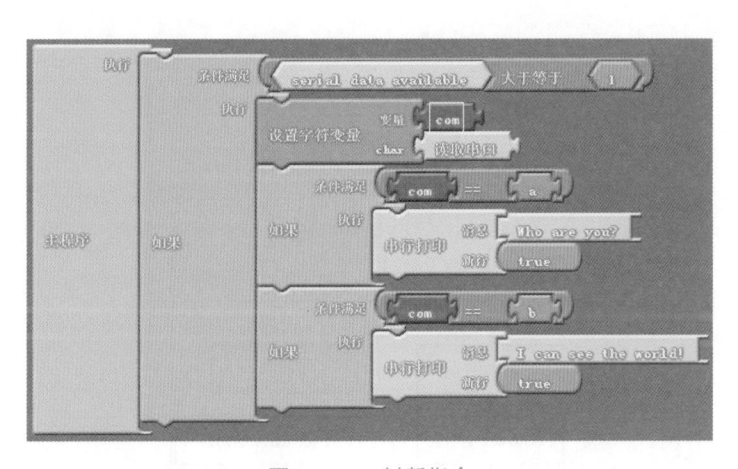

图2-1-6 判断指令

（a）

（b）

图2-1-7 指令输出界面

4. 蓝牙控制电动机运行

按照之前的程序，已经知道了蓝牙和控制器之间是如何进行信息传输的，那么接下来通过蓝牙来控制电动机的运行。

下面让6号引脚、9号引脚为高电平，5号引脚、10号引脚为低电平，两个电动机都是正转，如图 2-1-8所示。

让5号引脚、6号引脚、9号引脚、10号引脚都为低电平，两个电动机都不转动，如图2-1-9所示。

通过蓝牙指令"a"来控制两个电动机正转，通过蓝牙指令"b"来控制两个电动机停止，如图2-1-10所示。

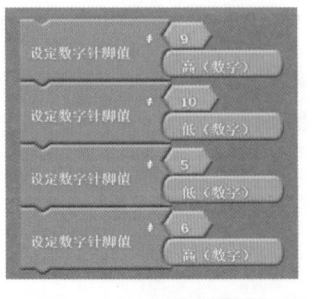

图2-1-8 电动机正转

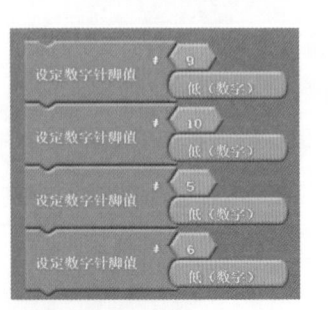

图2-1-9 电动机停止

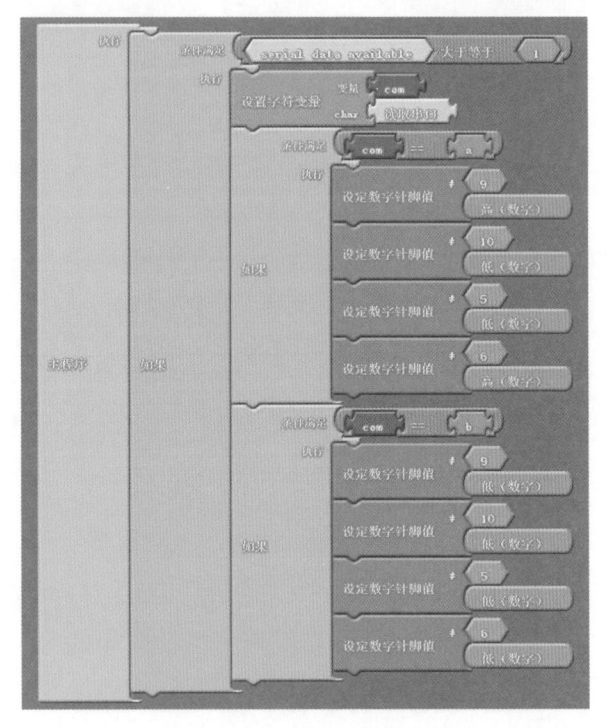

图2-1-10 蓝牙控制电动机

任务二 智能鼠眼睛和腿的协同工作

在红外调试实验中已经知道，当红外传感器检测到挡板时，输出低电平；当红外传感器检测不到挡板时，输出高电平。

（1）首先设置左前，右前传感器两个数字变量，如图2-2-11所示。

（2）红外传感器检测外界信息

图2-1-11 左前、右前红外传感器

后的输出值只有两种，即高电平或低电平，这和按键开关完全相同。可以将红外传感器当作开关来使用。

（3）设定红外发射端引脚，并且设置其发射频率。

（4）通过电动机的两种运行状态"转动"和"停止"来观察传感器输出的电平信号状态。

（5）整理组合前面的程序，如图2-2-12所示。

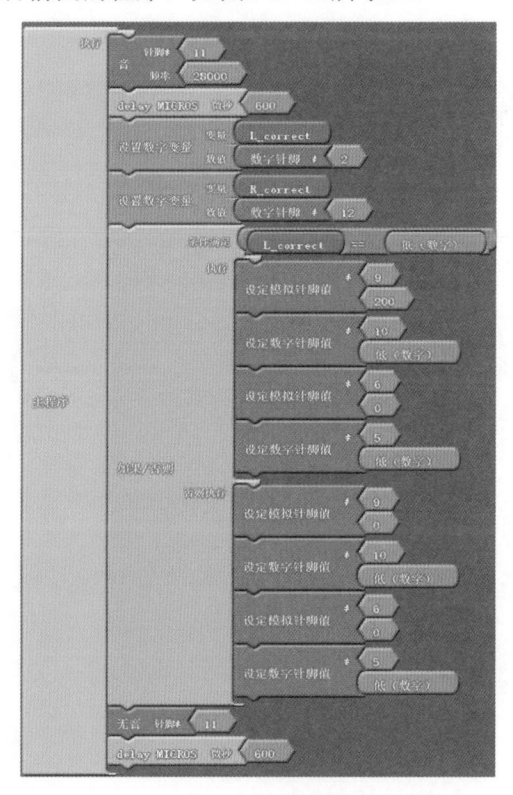

图2-1-12　红外传感器控制电动机动作

在设定好红外传感器发射强度后，开始下载程序，将智能鼠放置在开阔位置，小心遮挡左前方红外传感器验证结果。也可以将其他两组红外传感器加入到编程当中。

思考与总结

（1）除了蓝牙外，还有哪些常见的无线控制方式？

（2）当多个传感器同时工作时，需要注意相互之间的干扰问题；由于发射头使用相同的引脚，所以发射频率是相同的，可以采用交替发射检测的方式，提高检测精度。

项目二

智能鼠的姿态控制

学习目标

（1）学习智能鼠电动机速度的精确控制方法。

（2）学习如何实现智能鼠转弯。

智能鼠在迷宫中行走时，需要对电动机的速度进行精确控制，才能实现无碰触的直行和转弯。

任务一　智能鼠跑起来

在电动机驱动部分已经学习过如何使电动机转动。下面通过一系列操作来实现智能鼠直行。

在实际当中，由于电路、导线以及轮胎的原因，两个电动机的转速是不可能完全相同的，随着转动距离的增加，左右两个电动机的差速也会逐渐增大，导致智能鼠运行的姿态发生偏移。

一、脉宽调制（PWM）

1. PWM原理简介

脉宽调制是一种模拟控制方波宽度调制，是一种对模拟信号电平进行数字编码的方法。PWM是利用微处理器的数字输出来对模拟电路进行控制的一种非常有效的技术，广泛应用在从测量、通信到功率控制与变换的许多领域中。通过高分辨率计数器的使用，方波的占空比被调制用来对一个具体模拟信号的电平进行编码。PWM信号仍然是数字的，因为在给定的任何时刻，满幅值的直流供电要么完全有（ON），要么完全无（OFF）。电压源或电流源是以一种通（ON）或断（OFF）的重复脉冲序列被加到模拟负载上去的。通的时候即直流供电被加到负载上的时候，断的时候即供电被断开的时候。只要带宽足够，任何模拟值都可以使用PWM进行编码。

PWM就是脉宽调制，也就是占空比可变的脉冲波形，如图2-2-1所示。对半导体开关器件的导通和关断进行控制，使输出端得到一系列幅值相等而宽度不相等的脉冲，用这些脉冲来代替正弦波或其他所需要的波形。按一定的规则对各脉冲的宽度进行调制，既可改变逆变电路输出电压的大小，也可改变输出频率。

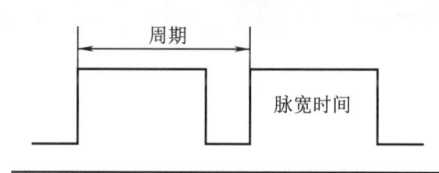

图2-2-1　脉宽调制

2. PWM和占空比的概念

（1）PWM：又称脉冲宽度调制技术，是一种模拟控制方式。PWM波形如图2-2-2所示。

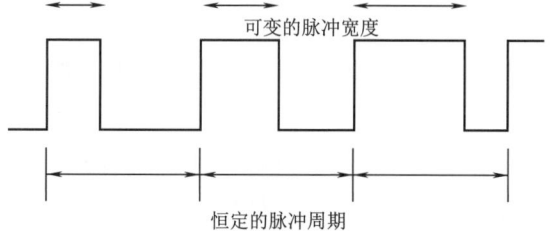

图2-2-2　PWM波形

（2）占空比：指高电平在一个周期中所占的比例。例如，占空比为50%，即高电平占整个周期时间的一半。

3. PWM定频调速原理

在PWM调速系统中，一般可以采用定宽调频、调宽调频、定频调宽三种方法改变控制脉冲的占空比，但是前两种方法在调速时改变了控制脉宽的周期，从而引起控制脉冲频率的改变。当该频率与系统的固有频率接近时将会引起振荡。为避免这种现象，采用定频调宽改变占空比的方法来调节直流电动机电枢两端的电压。

在调节速度时，定频调宽又称定频调速，这是在脉冲波形的频率不变的前提下（脉冲波形的周期不变），通过改变一个周期波形中高电平的时间从而改变波形的占空比，从而改变平均电压，调整电动机的转速。假定电动机始

终接通电源时，电动机最大转速为v_{max}，占空比为$D=t/T$, 则电动机的平均速度$v_d=D×v_{max}$，由公式可知，当改变占空比D时，就可以得到不同的电动机平均速度v_d，从而达到调速的目的。

二、智能鼠无校正走直线

编写智能鼠驱动程序，对引脚9、引脚10、引脚5、引脚6分别赋值高低电平，观察左、右电动机的转动方向，由此可以得到智能鼠直行程序，如图2-2-3所示。

将程序下载到智能鼠中，启动智能鼠观察偏移情况。

程序上载完成后，观察智能鼠的运行情况，智能鼠是否走直线，偏移较小？如果偏移较大应该怎么解决？

实际上，由于电路、导线以及轮胎的原因，两个轮子的转速是不可能完全相同的，也就是说肯定会发生或大、或小的偏移。那应该怎么解决呢？

回忆前面做过的电动机转速控制实验，即改变电动机引脚的模拟量大小即可实现对电动机转速的控制。所以将引脚6、引脚9替换为模拟量，并设置蓝牙调节来改变引脚6、引脚9模拟量的大小，最终实现智能鼠走直线或者在允许的范围内较小的偏移。

（1）首先设置两个变量"1_motor"、"r_motor"，并分别赋值150，如图2-2-4所示。

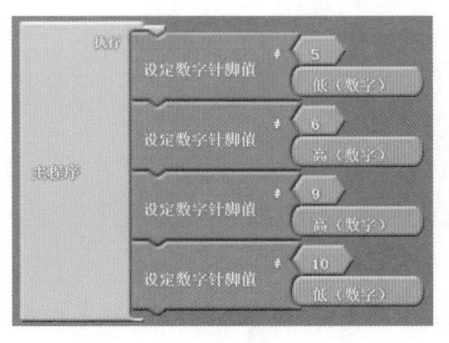

图2-2-3　智能鼠直行程序　　　图2-2-4　左右电动机模拟量分别赋值程序

（2）为了让程序更易理解，引入子程序模块。在"控制栏"取出子程序（母子模块各两个），对蓝牙控制部分和温度控制部分分别进行处理。为了保证子程序母子模块的对应性，两者名字必须一致，如图2-2-5所示。

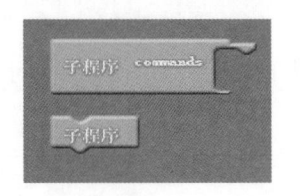

图2-2-5　左子程序模块（母子模块）

（3）对前进和停止分别应用子程序模块，如图2-2-6、图2-2-7所示。

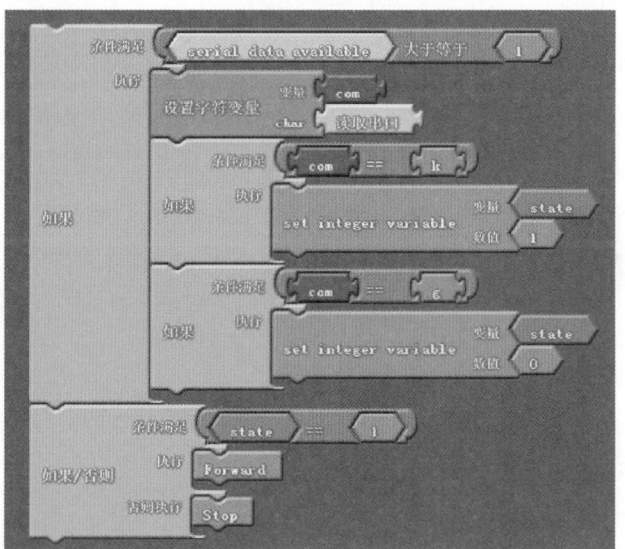

图2-2-6　前进子模块　　　　　　　　图2-2-7　停止子模块

这样，在主程序中只需使用 Forward 和 Stop 即可完成对智能鼠启动和停止的程序调用。

（4）现在继续加入蓝牙控制智能鼠启动和停止的部分，如图2-2-8所示。

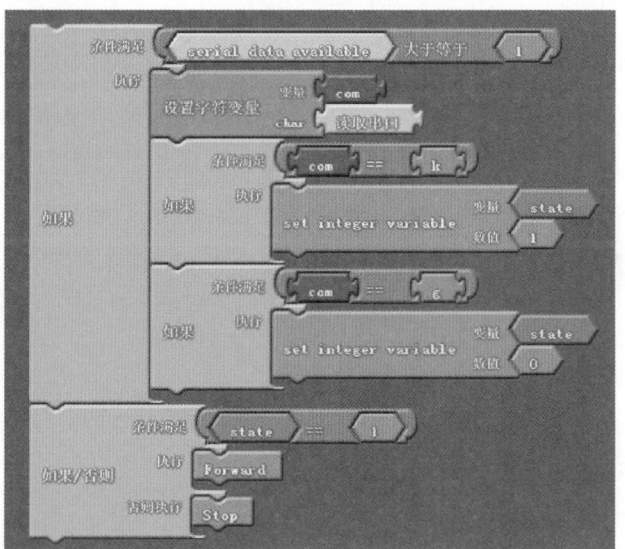

图2-2-8　手机蓝牙控制智能鼠启动和停止的程序

当手机蓝牙发送"k"时，变量"state"值为1，智能鼠开始运行；当手机蓝牙发送"g"时，变量"state"值为0，智能鼠停止运行。

（5）通过对"l_motor"、"r_motor"两个变量自加、自减来实现电动机转速的改变，如图2-2-9所示。

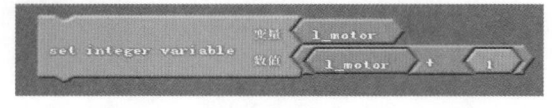

图2-2-9　实现电动机转速的改变

图2-2-9中，加号可以更改成减号，自加自减的数值可大可小。右电动机"r_motor"同理。

（6）将变量加减的程序扩展一下，并加入手机蓝牙控制，如图2-2-10所示。

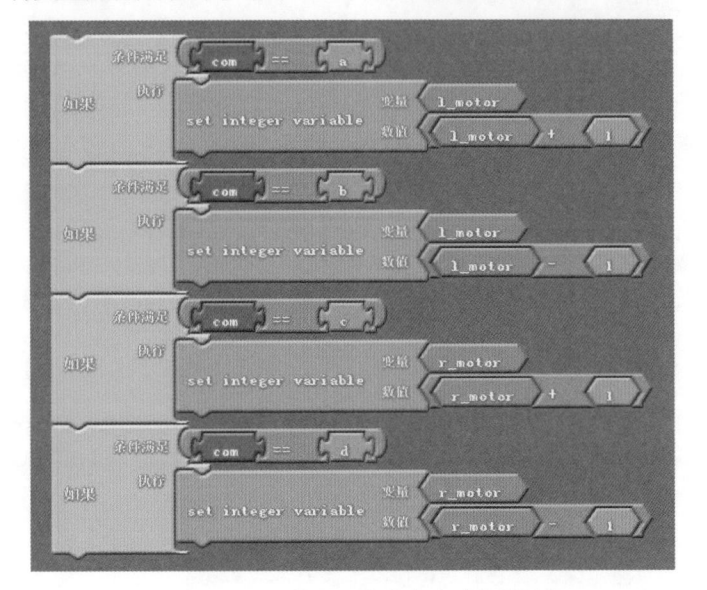

图2-2-10 蓝牙调节左右电动机转速

由图2-2-10可见，当蓝牙发送"a"时，左电动机加一个模拟量的速度；蓝牙发送"b"时，左电动机减一个模拟量的速度。当蓝牙发送"c"时，右电动机加一个模拟量的速度；蓝牙发送"d"时，右电动机减一个模拟量的速度。

（7）最后添加"通信"栏中的串行打印功能，并将两个电动机的速度发送到手机App上，记录下来留待以后实验使用，如图2-2-11所示。

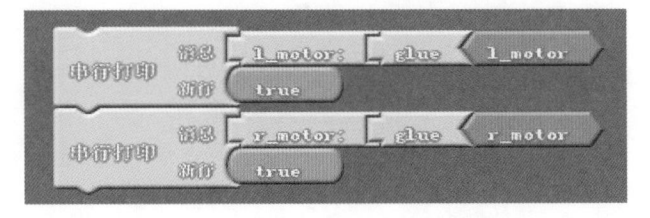

图2-2-11 串行打印左右电动机转速

由于蓝牙发送数据很快，所以需要添加一个延时函数来帮助我们获取数据。选用"控制"栏中的"延时"模块。

（8）到此为止，整个实验的所有功能已经全部实现了，总的程序如图2-2-12所示。

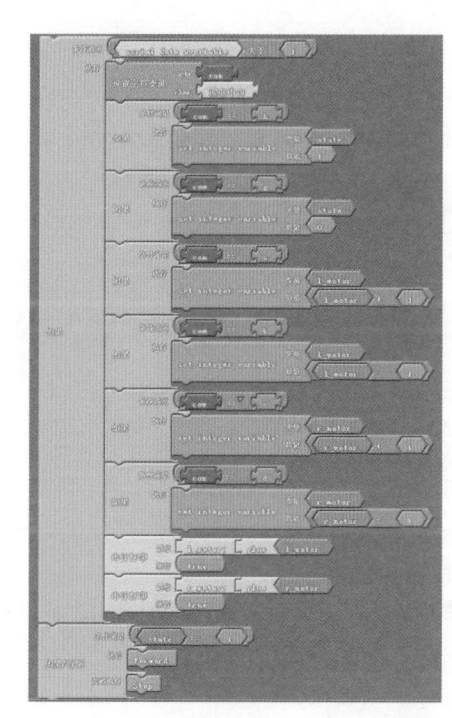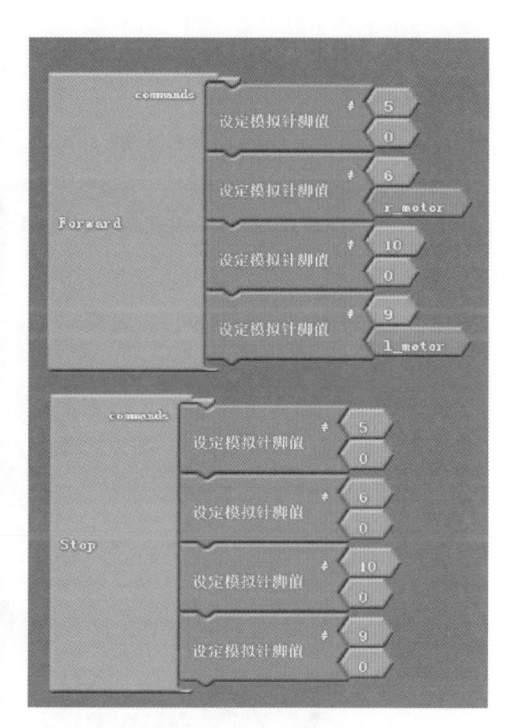

（a）蓝牙控制智能鼠走直线主程序模块　　　（b）蓝牙控制智能鼠走直线子程序模块

图2-2-12　蓝牙控制智能鼠走直线程序

三、智能鼠循迹走直线流程图（见图2-2-13）

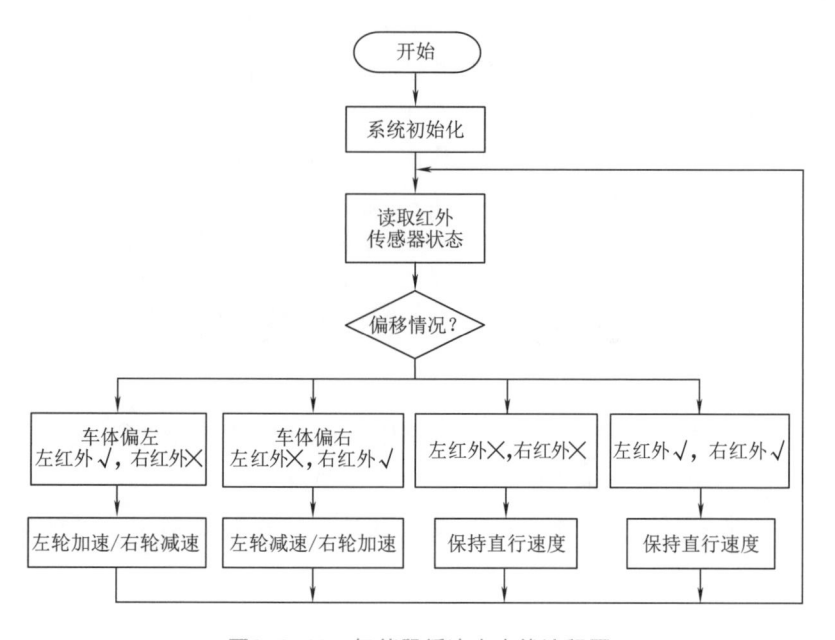

图2-2-13　智能鼠循迹走直线流程图

四、图形化编程

下面继续学习如何使用红外传感器来自动校正车姿。

结合红外调试部分的红外发射强度调试实验，可知：

当智能鼠左偏时，变量L_correct为低电平，R_correct为高电平。

当智能鼠右偏时，变量R_correct为低电平，L_correct为高电平。

所以，可以依据这两个结果对电动机进行调速。

假设测得智能鼠无校正走直线的速度为左电动机200、右电动机180。当智能鼠左偏时可以减小右电动机速度使智能鼠回到中心线上；当智能鼠右偏时可以减小左电动机速度使智能鼠回到中心线上。假设减小的模拟量为20，编程如图2-2-14所示。

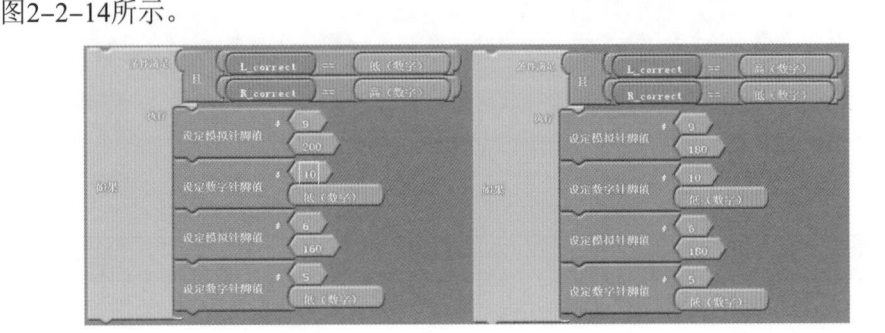

图2-2-14　智能鼠偏左、偏右校正

智能鼠竞赛迷宫大小为1.5 m×1.5 m，每个位置的光照情况都不相同，所以为了程序的完整性以及意外情况的发生，添加左前和右前传感器同时检测到挡板和同时检测不到挡板的情况，智能鼠左右电动机速不变，如图2-2-15所示。

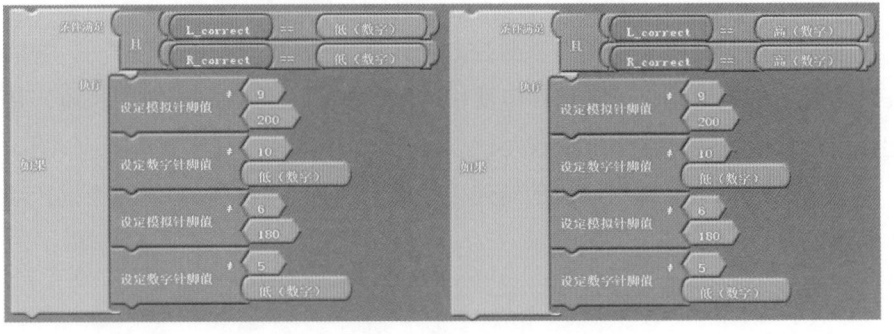

图2-2-15　智能鼠特殊情况不调速

为了提高程序的可读性，设置子程序：

Left_offset——左偏移子程序；

Right_offset——右偏移子程序；

视 频

智能鼠走
直线

Forward——前行子程序，如图2-2-16所示。

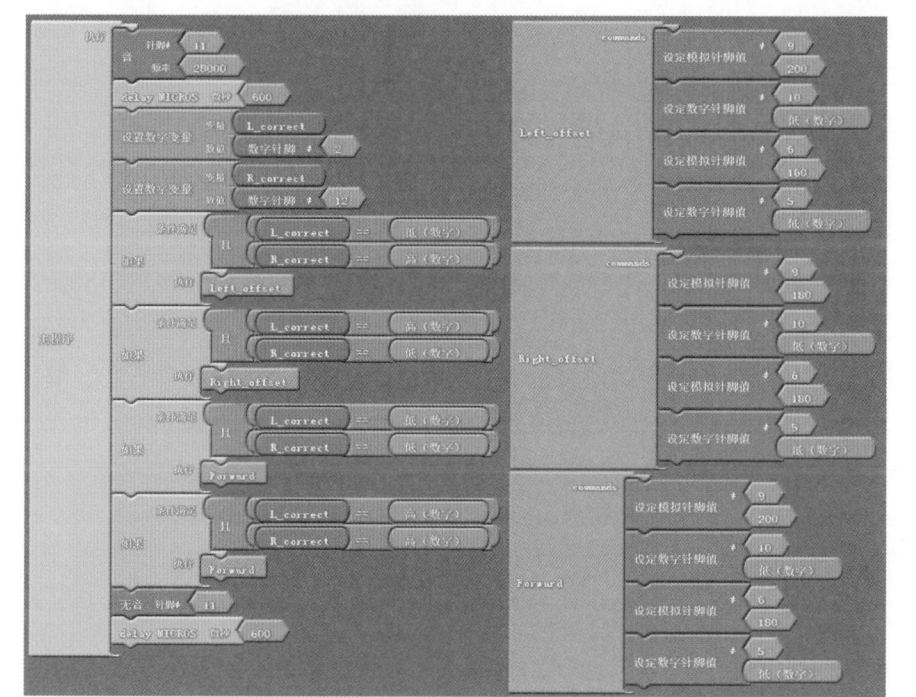

图2-2-16　有校正走直线

现在可以下载程序，然后将智能鼠放在测试场地的③和④号位置，验证智能鼠的自我校正了。

任务二　智能鼠学会转弯

智能鼠转弯角度是由两个参数控制的，即电动机转速和转弯时间。智能鼠在转弯时，外侧电动机的转速和电动机转动的时间（即延时函数的长短）共同起作用，数值越大，转弯角度越大。

通过匹配调节外侧电动机的转速以及转弯时间，就可以实现智能鼠转动任意角度。

一、转弯90°流程图

智能鼠在迷宫中行走时，基本的转弯角度是90°，判断智能鼠转弯是否90°的流程图如图2-2-17所示。

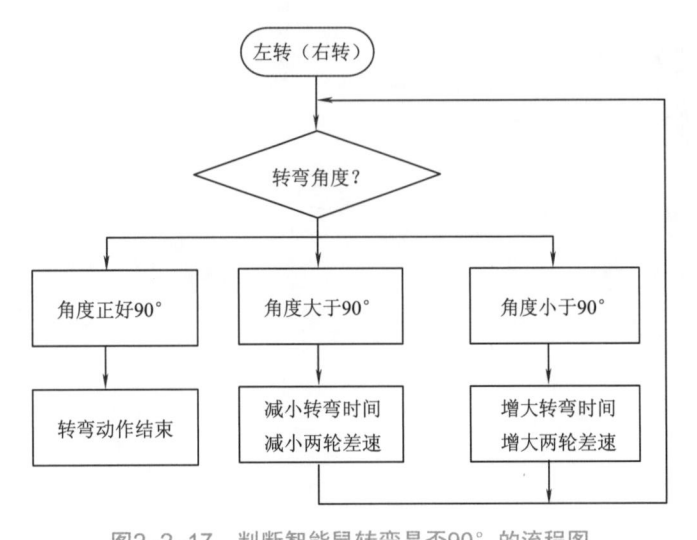

图2-2-17　判断智能鼠转弯是否90°的流程图

通过测试场地上的⑪、⑫和⑬标号，可以判断智能鼠是否准确转弯90°。假设发现智能鼠左转不是90°，此时有两种调节方法（以左转为例）：

（1）增加或减小引脚6模拟值大小。

（2）增加或减小左转时间。

二、精确控制转弯角度图形化编程

学会了如何判断和调整转弯角度，下面进行实际程序编写。

（1）由于需要对速度和时间进行调整，所以设置速度和时间为变量，并赋初值，如图2-2-18所示。

（2）加入蓝牙控制智能鼠的启动和停止。如果直接采用发送指令控制智能鼠停止，会引入人为误差，造成一定的干扰，所以改变策略：设置一个系统状态变量来表示智能鼠是运行状态还是停止状态，如图2-2-19、图2-2-20所示。

视频 ●

智能鼠精确转弯

图2-2-18　对速度和时间设置变量程序

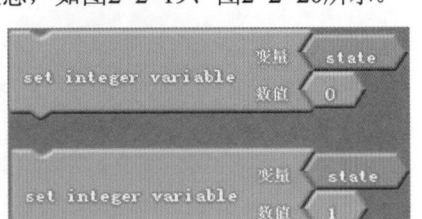

图2-2-19　设置状态比较

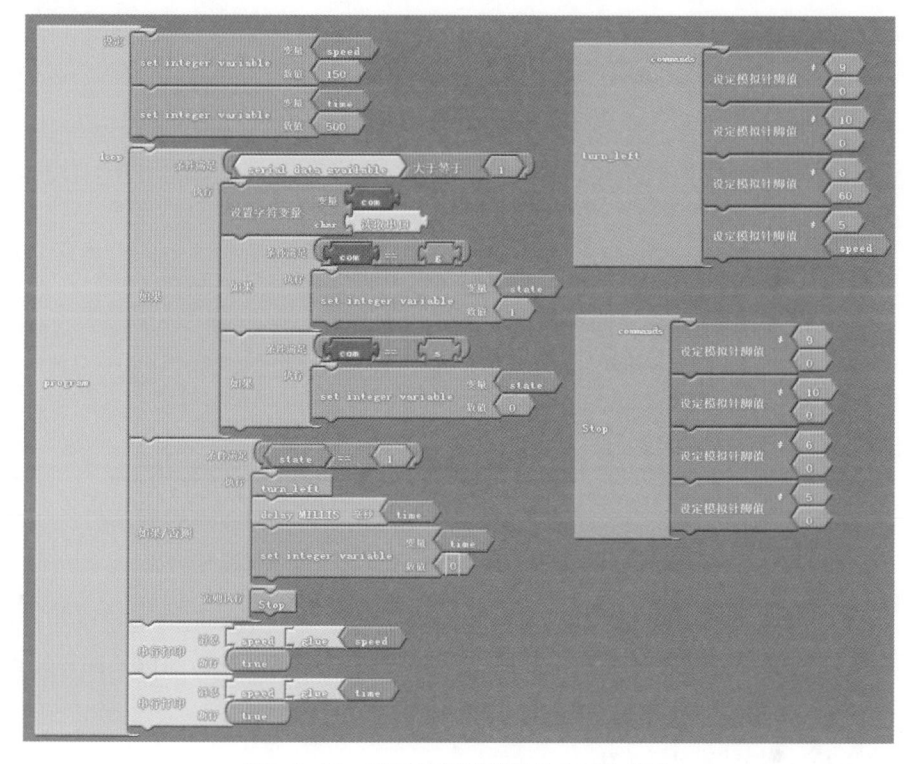

图2-2-20 蓝牙控制智能鼠的启动和停止

接下来就是对智能鼠转弯90°的控制。将Forward子程序替换为左转弯子程序，如图2-2-21所示。

如果初值不能满足智能鼠恰好转弯90°，应该怎么办呢？具体做法就是通过蓝牙发送指令，自加或自减改变变量"speed"或"time"的大小，如图2-2-22所示。

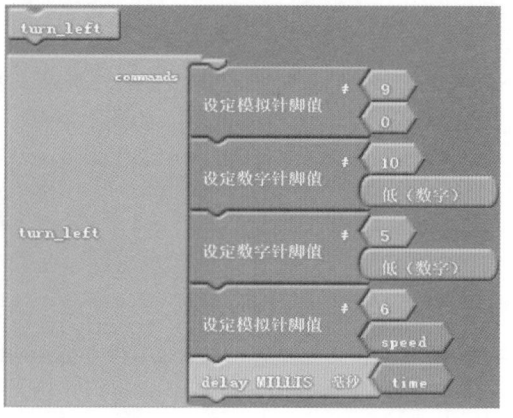

图2-2-21 左转弯子程序

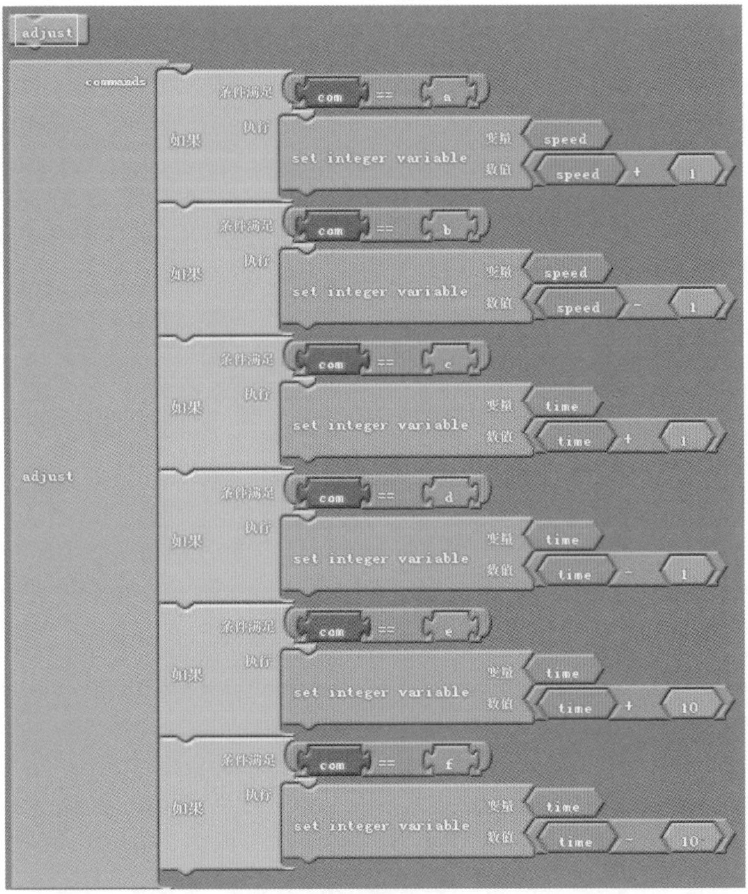

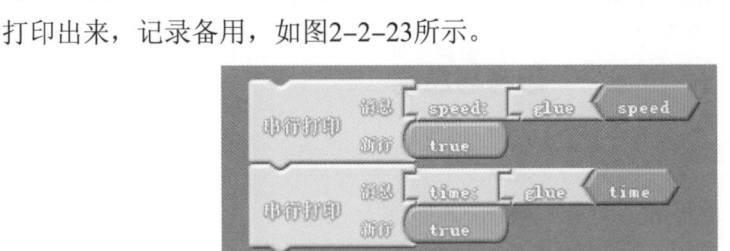

图2-2-22 蓝牙控制速度和转弯时间自加、自减

通过蓝牙发送"a""b""c""d""e""f"时，速度和时间进行自加、自减，来求得准确的速度和时间。最后，不要忘记将速度和时间通过串行打印出来，记录备用，如图2-2-23所示。

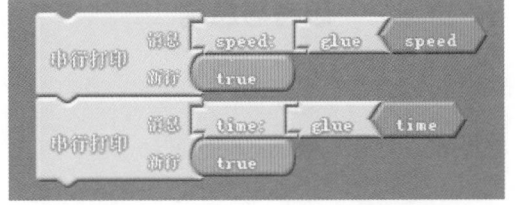

图2-2-23 串行打印速度和时间

由于智能鼠转弯中存在延时函数，所以串行打印后不需要再添加延时函数。至此，就完成了智能鼠转弯90°所需的速度和时间的求解了。

将程序下载到智能鼠中，发送蓝牙指令观察智能鼠的动作，如图2-2-24所示。

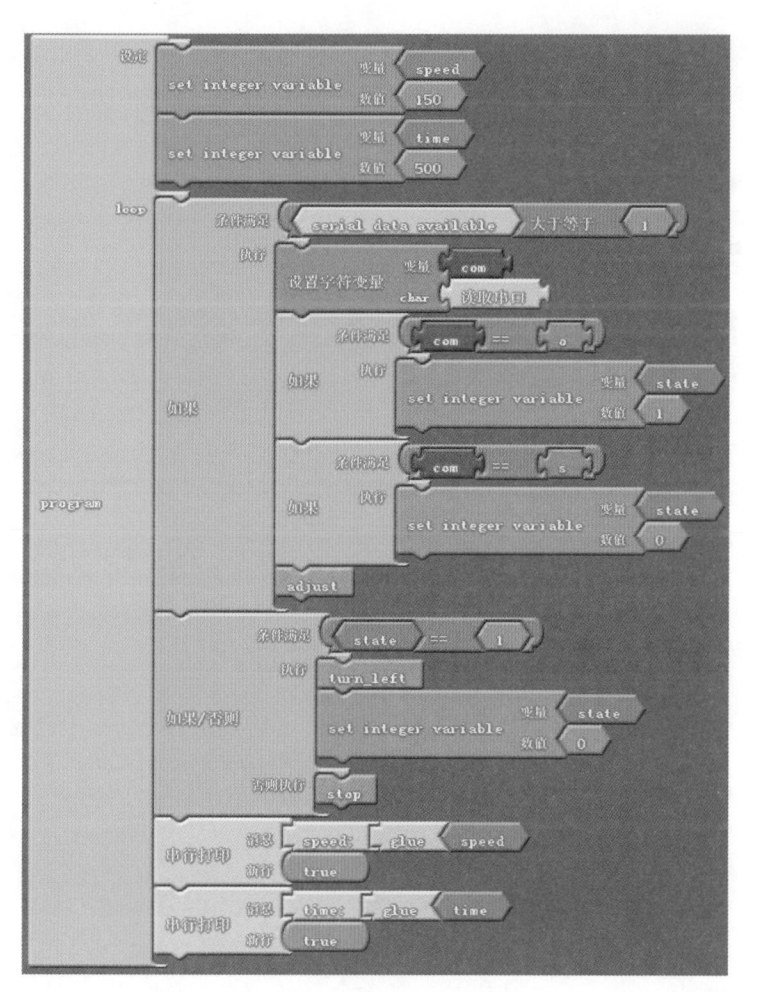

图2-2-24　智能鼠的连续动作（前进，停，左转90°，停，前进，停）程序

不要忘记将测得的速度和时间记录下来，以备在后面的实验中使用。

思考与总结

（1）智能鼠的转弯方式有哪些？相比较而言，各有什么优缺点？

（2）电动机控制精度决定了智能鼠的成绩。

第二篇　竞赛创新篇

　　前两篇中已经介绍了智能鼠的软硬件以及智能鼠的基础编程调试方法，针对中国IEEE国际标准Micromouse走迷宫竞赛的要求，本篇主要介绍智能鼠优化算法。掌握智能鼠走迷宫竞赛的规范，可以最快的速度完成迷宫搜索和最优路径选择；进行以往智能鼠竞赛迷宫案例要点分析，为参加中国IEEE国际标准Micromouse走迷宫竞赛做好准备。

项目一

智能鼠常用算法解析

学习目标

学习智能鼠是如何智能选择路径的。

如何才能让智能鼠在迷宫中快速漫游呢？智能鼠的主要任务是根据IEEE国际标准Micromouse走迷宫竞赛规则完成迷宫搜索和最优路径选择的，是考察一个系统对一个未知环境的探测、分析及决策能力的一种比赛，下面来简单了解一下这方面的知识。

在没有预知迷宫路径的情况下，智能鼠必须要先探索迷宫中的所有单元格，直到抵达终点为止。做这个处理的智能鼠要随时知道自己的位置及姿势，同时要记录下所有访问过的方块四周是否有挡板。在搜索过程中为了节约搜索时间，还要尽量避免重复搜索已经搜索过的地方。

那么，怎样来探索迷宫呢？通常有两种策略：（1）尽快到达目标地；（2）搜索整个迷宫。

这两种策略各有利弊。利用第一种策略虽然可以缩短探索迷宫所需的时间，但是不一定能够得到整个迷宫的地图资料。若找到的路不是迷宫的最优路径，这将会影响智能鼠最后冲刺的时间。利用第二种策略，可以得到整个迷宫的地图资料，这样就可以求出最优路径。不过采用这种策略所使用的搜索时间较长。

任务一　左手法则

要想完成智能鼠走迷宫竞赛，必须知道如何把迷宫搜索的基本方法用程序编写出来，下面首先介绍左手法则算法。

在迷宫搜索方法策略上，智能鼠优先考虑向左转弯，其次是向前直行，最后考虑向右转弯，这种策略称为左手法则。如图3-1-1所示，图中智能鼠出发点是坐标（0，0），虚线为智能鼠运动路径，可以很清楚地看到，每当智能

视频

智能鼠算法
解析(左手)

鼠遇到分支路口时，它都会选择优先向左转弯，不能左转弯时智能鼠会选择直行，当智能鼠既不能左转弯也不能直行时才会向右转弯。

通过程序设定的方式，使智能鼠遇到路口时优先进行左转弯，如图3-1-2所示。

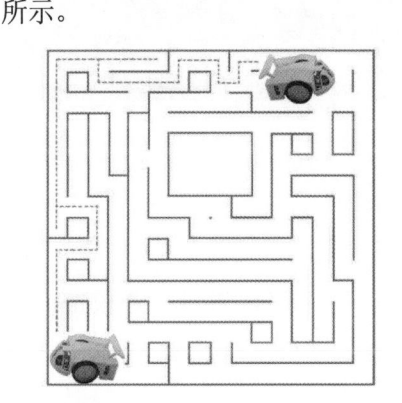
图3-1-1　左手法则示意图

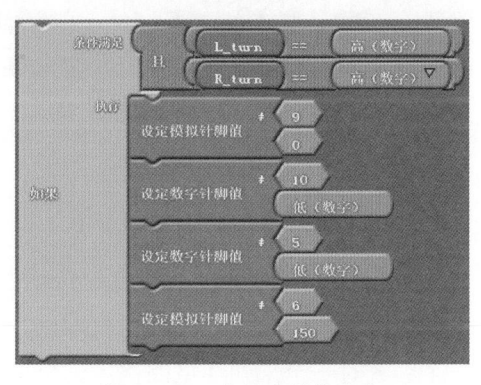
图3-1-2　左转弯优先程序

当后方两个传感器都没有检测到挡板时，左轮停止转动，右轮正向转动，进行左转弯。

以左手法则示意图中的几个关键点为例进行解释，决策对应表如表3-1-1所示。在坐标（2，6）处，智能鼠可以选择左转或右转，依据左手法则最终会选择左转；在坐标（1，8）处，智能鼠可以选择前进或右转，依据左手法则最终会选择前进；在坐标（2，F）处，智能鼠可以选择右转或前进，依据左手法则最终会选择右转。

表3-1-1　左手法则示意图关键点决策对应表

坐标点	方向选择项	最终策略选择
（2，6）	左转、右转	左转
（1，8）	前进、右转	前进
（2，F）	右转、前进	右转

任务二　右手法则

顾名思义，左手法则的对立面就是右手法则，即智能鼠顺着右边走，只要右边存在没有走过的入口则向右转弯。

与左手法则相似，当智能鼠在前进时，如果在前进的方向上存在两条和两条以上的支路时，它需要选择向哪个方向转弯，转弯的方向不同导致智能鼠的运动路径也不同。智能鼠优先考虑向右转弯，其次向前直行，最后才考虑向左

转弯，这种策略称为右手法则。如图3-1-3所示，图中坐标（0，0）依然为智能鼠的出发点，虚线依然为智能鼠运动路径，不同的是，每当智能鼠遇到分支路口时，它都会选择优先向右转弯，不能右转弯时智能鼠会选择向前直行，当它既不能右转弯也不能直行时才会向左转弯。

同左手法则一样，当后方左右两个传感器全都检测到有路口时，根据程序，左轮正向转动，后轮停止，进而使智能鼠向右转弯，如图3-1-4所示。

视　频

智能鼠算法
解析(右手)

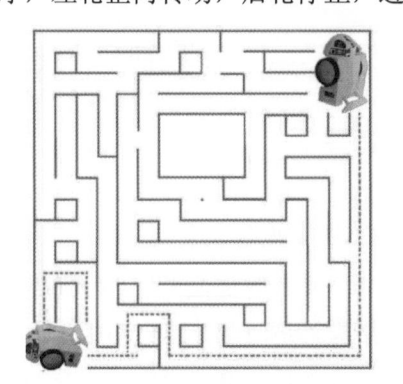

图3-1-3　右手法则示意图

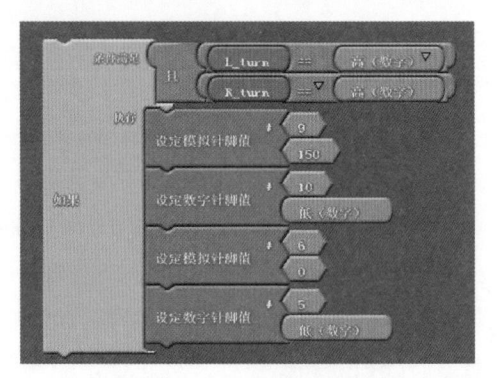

图3-1-4　右转弯优先程序

以右手法则示意图中的几个关键点为例进行解释，决策对应表如表3-1-2所示。在坐标（0，4）处，智能鼠可以选择前进或右转，依据右手法则最终会选择右转；在坐标（2，0）处，智能鼠可以选择左转或右转，依据右手法则最终会选择右转；在坐标（4，0）处，智能鼠可以选择左转或前进，依据右手法则最终会选择前进。

表3-1-2　右手法则示意图关键点决策对应表

坐标点	方向选择项	最终策略选择
（0，4）	前进、右转	右转
（2，0）	左转、右转	右转
（4，0）	左转、前进	前进

思考与总结

（1）单一使用左手法则和右手法则有何优缺点？

（2）左手法则和右手法则是依据转弯方向的优先级来设定的。当智能鼠位于不同的位置时选择不同的法则会显著提高智能鼠的搜索效率。

项目二

智能鼠高级功能

学习目标

（1）掌握多传感器协同工作的原理和控制方法。

（2）学习多线程工作的原理与实现方法。

（3）学习运用多传感器协同工作与多线程工作实现智能鼠走迷宫。

智能鼠融合多种技术、多种学科，是一个非常复杂的学习平台。外界感知的手段包括红外传感器检测与无线遥控，运动结构就是两台减速电动机。由于在竞赛时除启动和停止外，禁止使用无线遥控，所以智能鼠能否快速而准确到达终点，就取决于传感器的检测精度与电动机运行控制的准确度。本项目就通过两个任务的学习来分别提升传感器的检测精度与电动机控制的准确度，从而实现智能鼠快速而准确地走迷宫。

任务一　多传感器协同工作原理与实现

在前面"智能鼠看世界"的任务中，已经学习过如何实现传感器检测。鉴于接收头（IRM8601S）特性，红外发射器发射频率越接近38 kHz，检测距离越远；反之，检测距离越近。由此就可以实现迷宫挡板与路口的检测。

智能鼠左前、右前、左后与右后四组红外传感器分别检测对应方向有无挡板，需要使用的发射频率不尽相同。每组传感器的接收头使用独立的I/O接口，可以准确地接收对应方向的高低电平信息（有无挡板）；然而发射头使用公用的IO 11接口，使用某一个频率发射红外线时，极有可能造成其他三个方向检测结果出现错误，那么如何才能准确地实现四个方向的挡板检测呢？

下面学习使用传感器交替发射——多传感器协同工作的方法，来实现四组传感器使用各自频率发射红外线并检测对应方向的挡板信息。

一、流程图

设置变量IRNum表示红外发射顺序。取值不同时，发射不同的频率并调用

对应的红外传感器状态。红外交替发射流程图如图3-2-1所示。

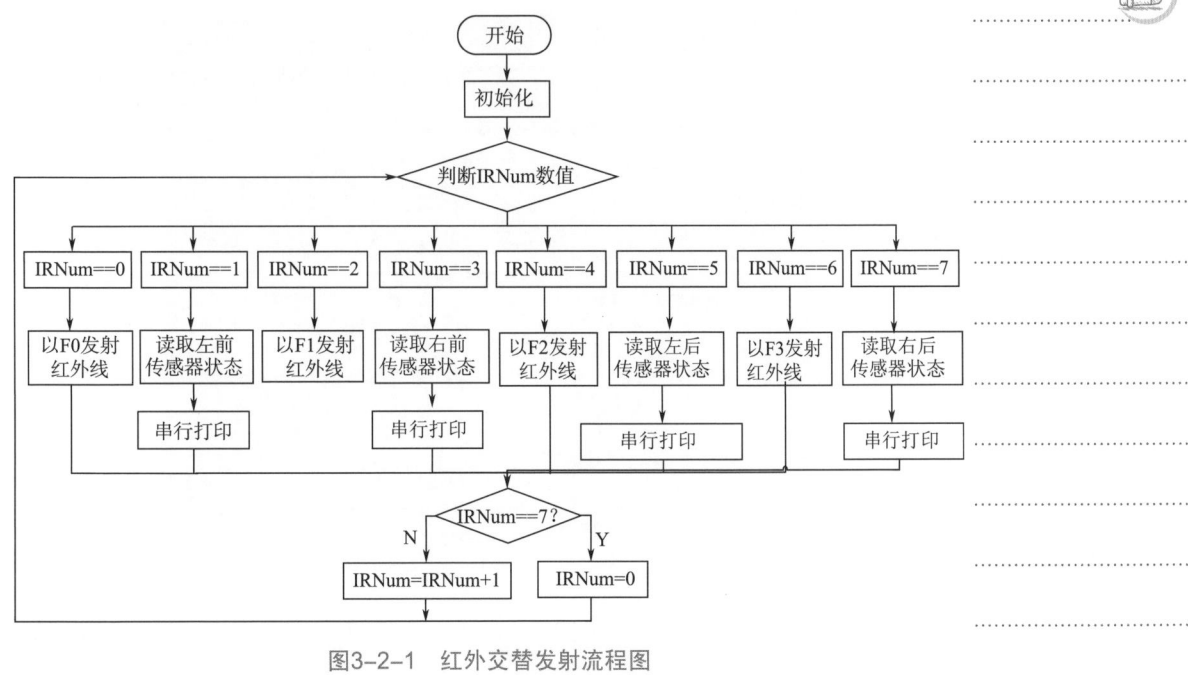

图3-2-1 红外交替发射流程图

二、图形化编程

（1）程序初始化。设置Data_init子函数表示程序数值化，其中变量IRNum取值范围为0~7，为发射顺序；F0~F3为左前、右前、左后和右后红外传感器的发射频率，如图3-2-2所示。

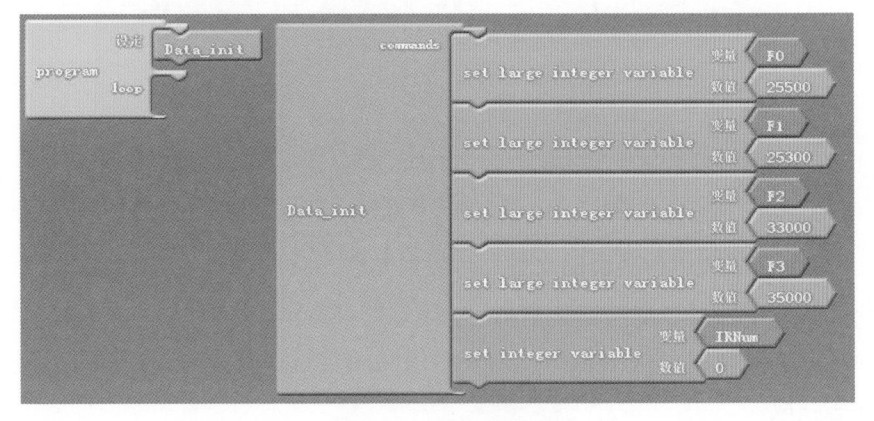

图3-2-2 程序初始化

（2）左前传感器发射与状态获取。主程序loop首次运行时变量IRNum被初始化为0，因此左前传感器使用F0发射红外线，在程序最后变量IRNum需要

自加1，从而方便第二次运行时，直接运行需要的程序，如图3-2-3所示。

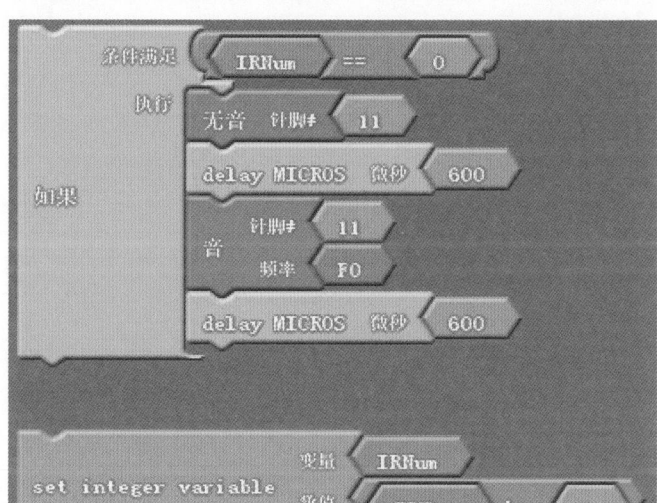

图3-2-3　第一次运行以自己的频率发射红外线

loop第二次运行时，变量IRNum=1，直接读取左前传感器的状态，如图3-2-4所示。

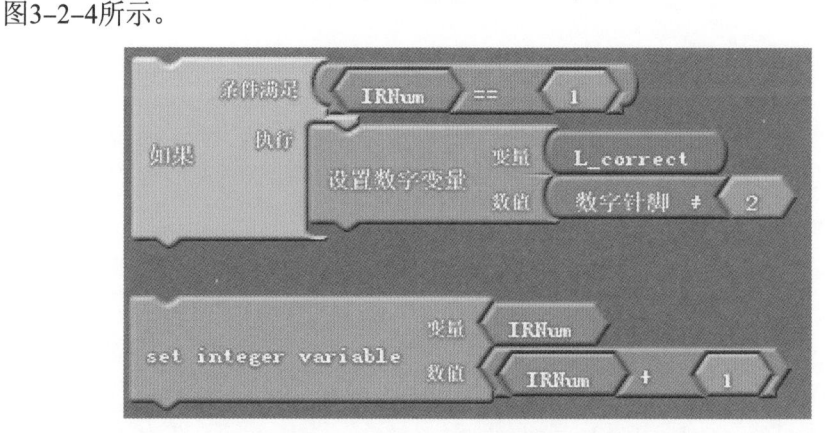

图3-2-4　第二次运行获取传感器状态

（3）完成四组传感器的交替循环发射与状态获取。按照（2）类推，完成四组传感器的交替发射与状态获取，如图3-2-5所示。需要注意，当完成最后一个的状态获取时，IRNum=7，需要对IRNum重新赋值为0，以完成循环发射与状态获取。

（4）组合程序，完成下载。为了方便观察结果，在每个传感器状态获取后添加"串行打印"图形，如图3-2-6所示。

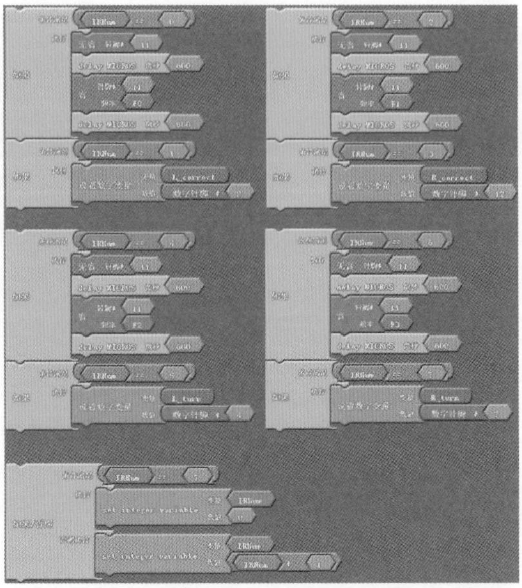

图3-2-5 四组传感器的交替发射与状态获取

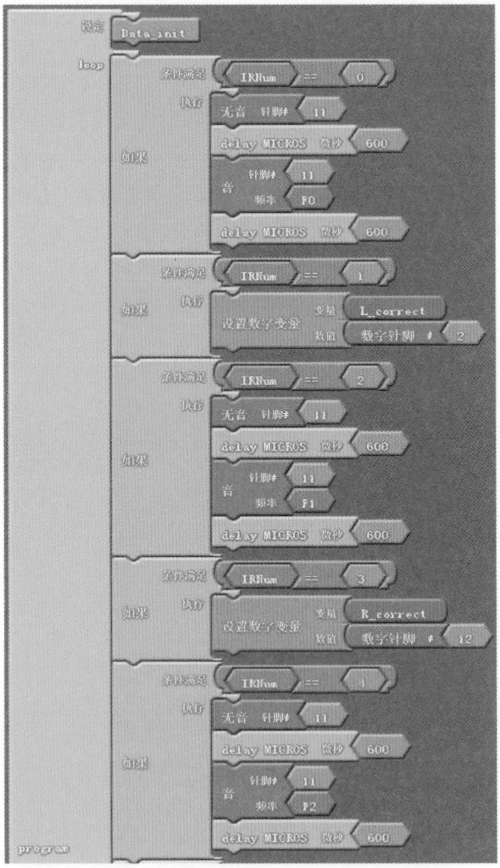

图3-2-6 智能鼠传感器协同工作全程序

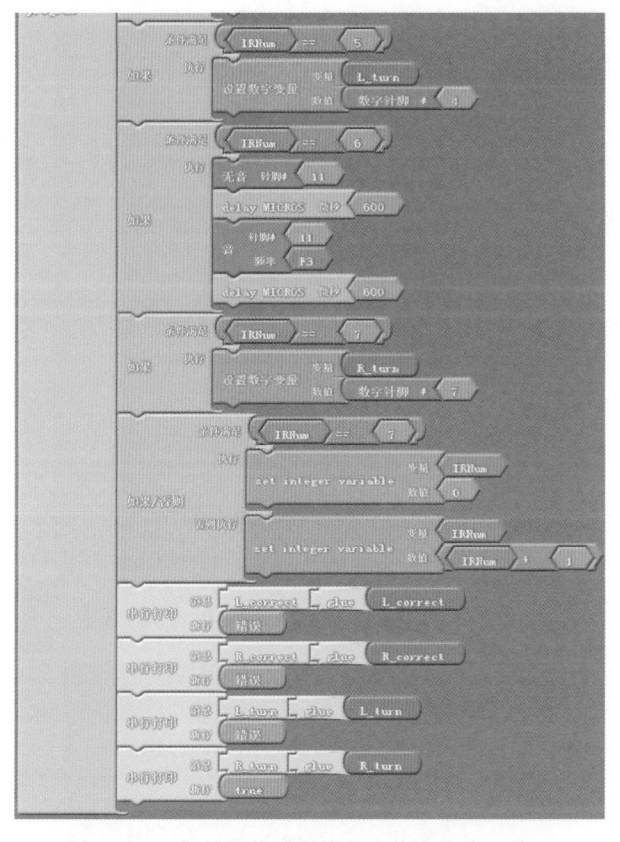

图3-2-6　智能鼠传感器协同工作全程序（续）

串口显示红外状态如图3-2-7所示。

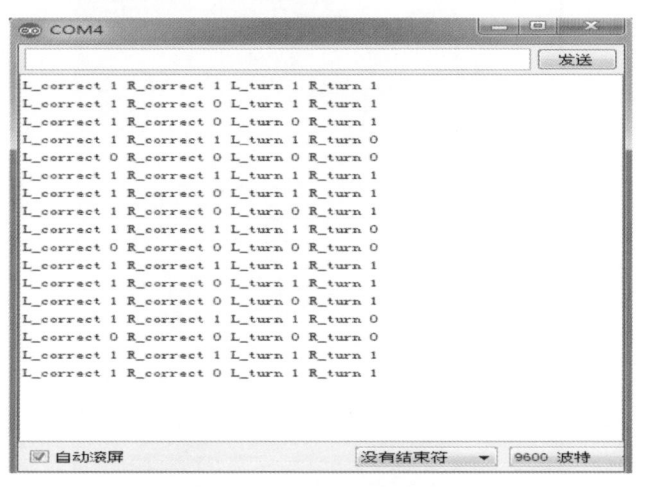

图3-2-7　串口显示红外状态

任务二　多线程工作原理与实现

在智能鼠的基础功能调试项目与姿态控制项目中已经学习过电动机的驱动、调速和转弯。下面来思考一个小实验：

左侧电动机正转0.5 s，停止0.3 s并循环；右侧电动机正转0.4 s，停止0.4 s并循环。

按照一贯的思路，这个实验能否完成呢？下面来尝试一下。

实验一　左右电动机的不同控制

1.流程图

左右电动机的不同控制流程图如图3-2-8所示。

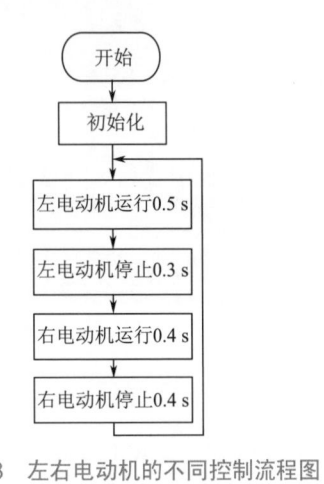

图3-2-8　左右电动机的不同控制流程图

2.图形化编程

（1）程序初始化。左右电动机的正转与停止如图3-2-9所示。

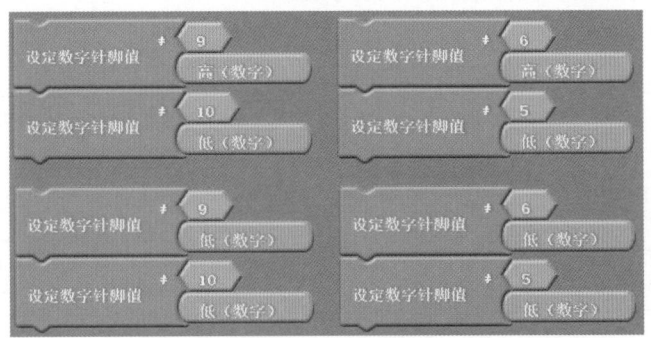

图3-2-9　程序初始化

（2）添加延时。分别添加延时，并命名为子程序left_motor与right_motor，如图3-2-10所示。

（3）组合并下载程序（见图3-2-11）。观察智能鼠电动机转动情况，结果并没有达到实验要求。

delay：保持控制器当前输出状态不变，等待一定的时间后继续执行后面的程序。

程序是从上而下顺序执行的，delay延时函数会对整个系统生效的，会造成之后的代码滞后执行。主程序过长也会造成系统的实时性变差，从而使电动机的

控制变得困难。下面介绍一种多线程的控制方式，来提高电动机的控制精度。

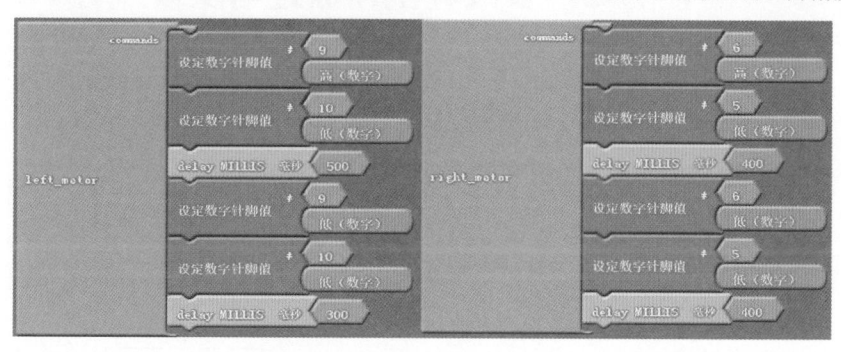

图3-2-10　命名电动机子程序

实验二　多线程控制电动机运行

多线程（multithreading），是指从软件或者硬件上实现多个线程并发执行的技术。也可以简单理解为多个主程序同时执行，除变量调用外，多线程之间互不干扰。

1. 流程图

多线程运行流程图如图3-2-12所示。

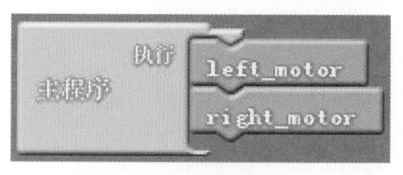

图3-2-11　组合程序

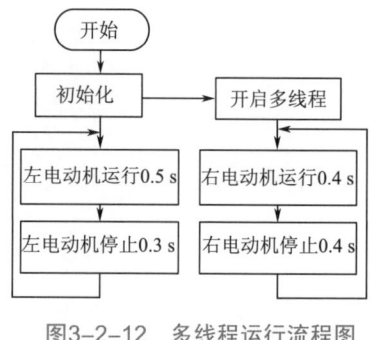

图3-2-12　多线程运行流程图

2. 图形化编程

多线程功能的实现，需要用到SCoop栏中的图形，如图3-2-13所示。

图3-2-13　多线程控制图形

SCoop中的Sleep延时只对多线程自身产生影响，对主函数不产生影响；同样的，delay延时对多线程不产生影响。

（1）初始化多线程并添加延时。将右电动机移动到多线程中，并替换延时为Sleep函数，如图3-2-14所示。

（2）组合并下载程序，如图3-2-15所示。

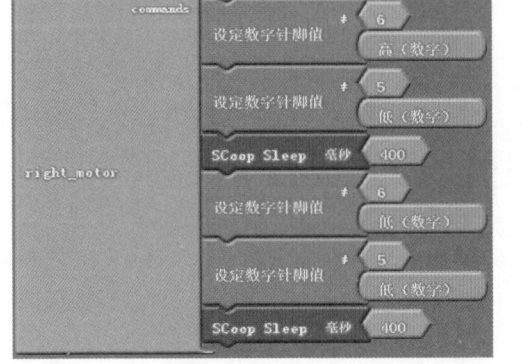 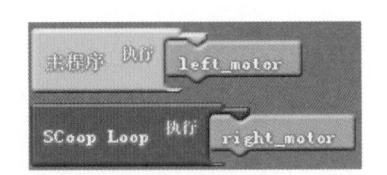

图3-2-14　为右电动机控制添加Sleep延时　　图3-2-15　主程序与多线程并行运行

运行程序可以发现，左右电动机的动作均达到了预期，互不干扰。

任务三　智能鼠躲避障碍灵活运行

本任务使用前面学过的提高传感器检测精度与电动机运行控制准确度的方法，来实现智能鼠快速而准确地走迷宫。

红外传感器发射与状态获取放在主程序中，电动机的控制（调速与转弯）放在多线程中。主程序与多线程并行运行，除变量调用外，互不干扰；程序的实时性得到提升。

1. 流程图

左前与右前传感器在运行时起校正车姿的作用，所以，

当左前传感器检测到挡板时，右电动机减速，反之保持高速；

当右前传感器检测到挡板时，左电动机减速，反之保持高速。

左后与右后传感器在运行时起路口检测的作用，所以：

当左后传感器检测到路口时，左转弯，否则直行；

当右后传感器检测到路口时，右转弯，否则直行。

注意： 在进入任意一个转弯时，都需要将多线程由校正走直线状态切换到转弯状态，在转弯结束后也要将多线程状态复位回校正走直线状态。

多线程流程图如图3-2-16、图3-2-17所示。

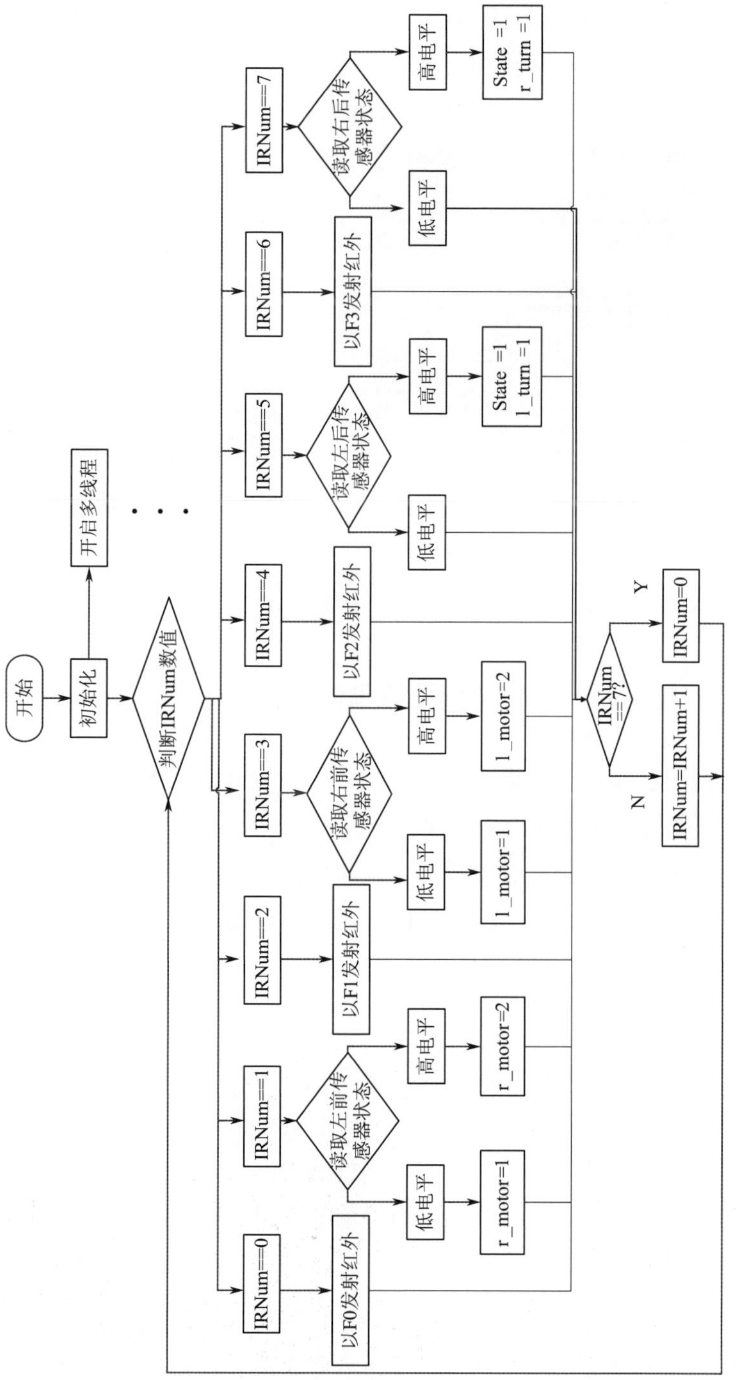

图3-2-16 多线程流程图（1）

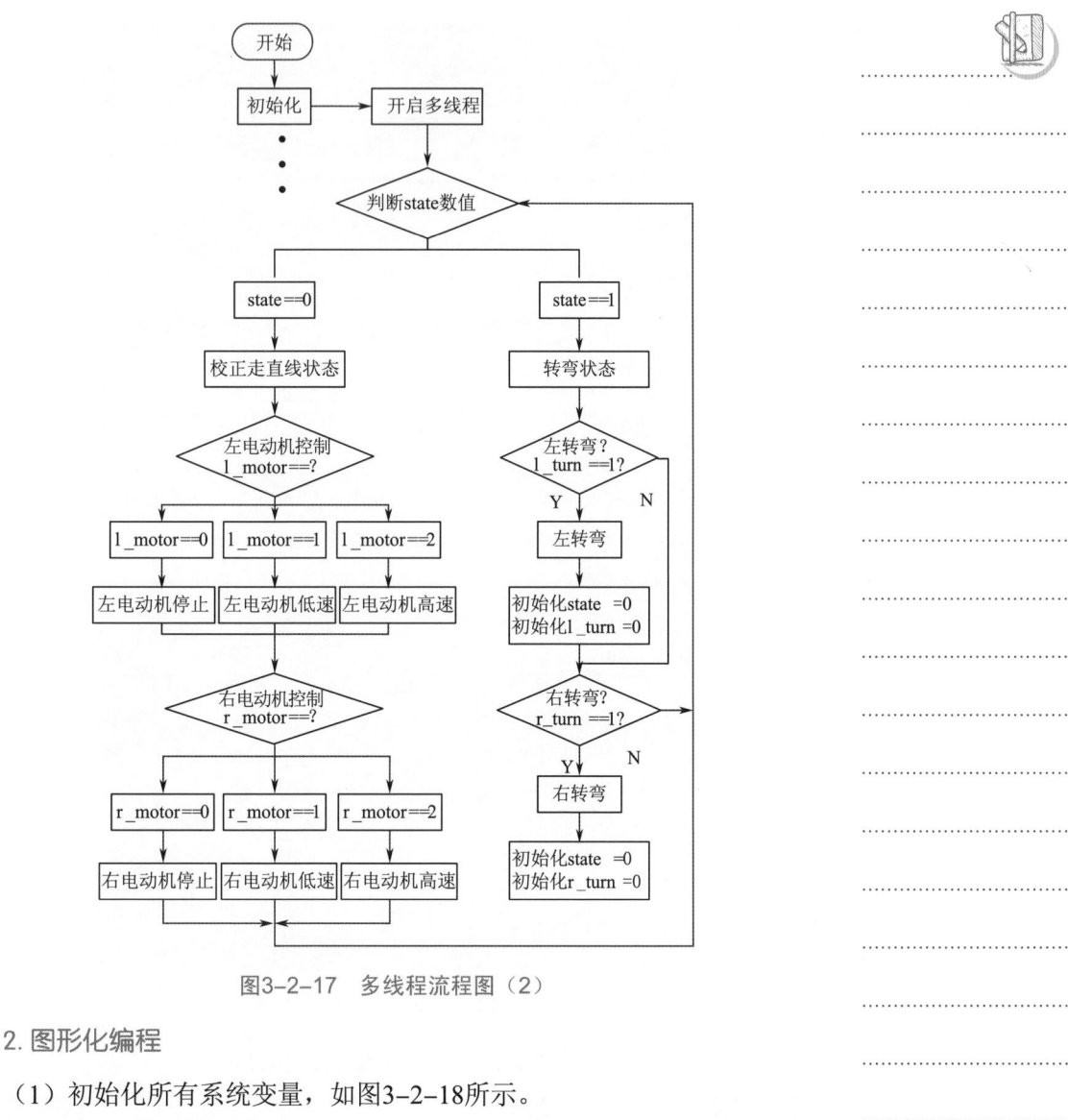

图3-2-17 多线程流程图（2）

2. 图形化编程

（1）初始化所有系统变量，如图3-2-18所示。

图3-2-18中的变量在程序运行时会进行调用或赋值，取值范围与含义如表3-2-1所示。

由于每只智能鼠的参数都是不同的，所以用户在调试时一定要替换为自己测得的数据。

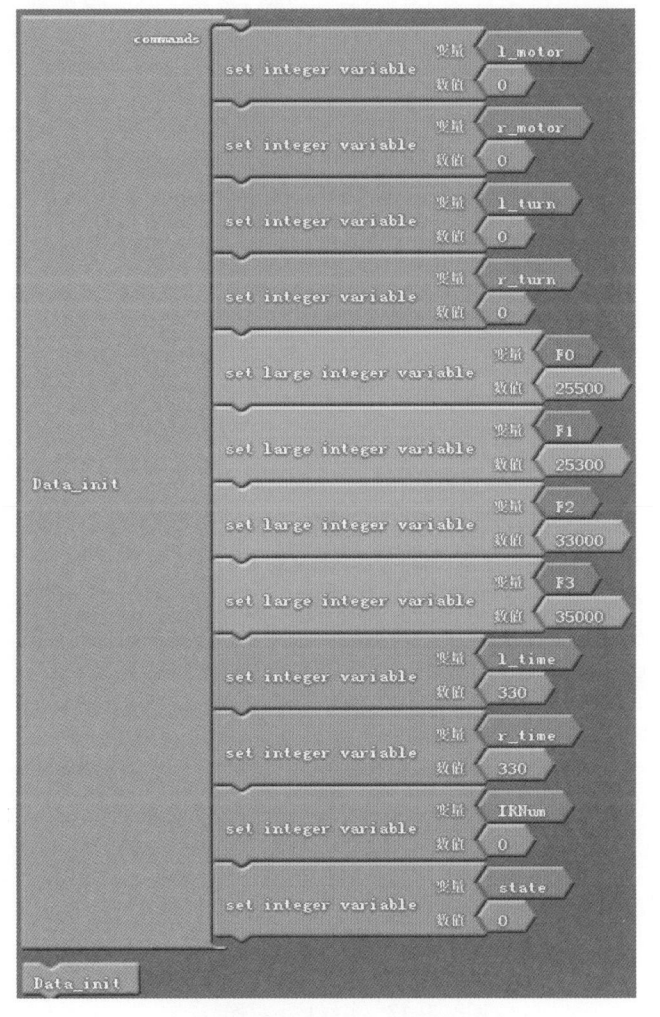

图3-2-18　系统初始化

表3-2-1　变量取值范围对照表

项目	含义	取值范围		
		0	1	2
r_motor	右电动机控制	停止	低速	高速
l_motor	左电动机控制	停止	低速	高速
state	多线程状态	校正走直线	转弯	无
l_turn	左转弯控制	直行	左转弯	无
r_turn	右转弯控制	直行	右转弯	无
F0~F3	四组发射频率	智能鼠的眼睛看世界任务获得		
l_time	左转90°所需时间	智能鼠学会转弯任务获得		
r_time	右转90°所需时间	智能鼠学会转弯任务获得		

（2）调用多传感器协同工作程序并加入变量赋值，如图3-2-19~图3-2-22所示。

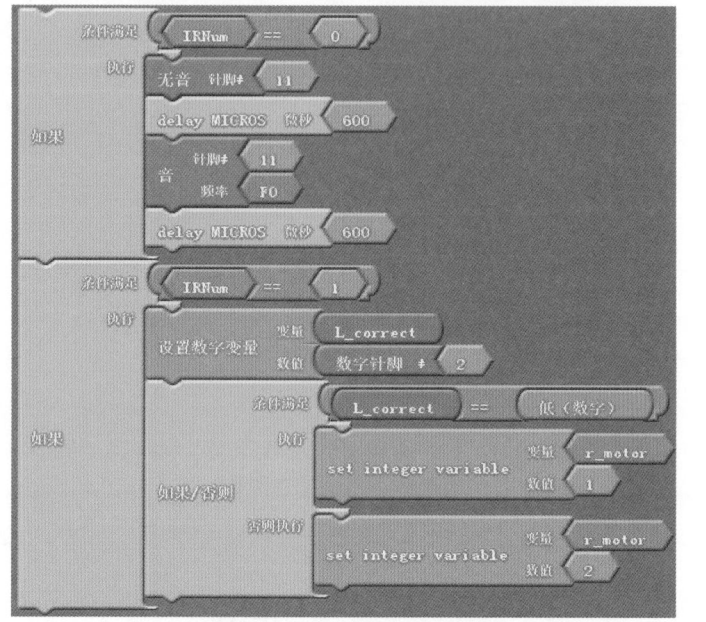

图3-2-19 左前传感器

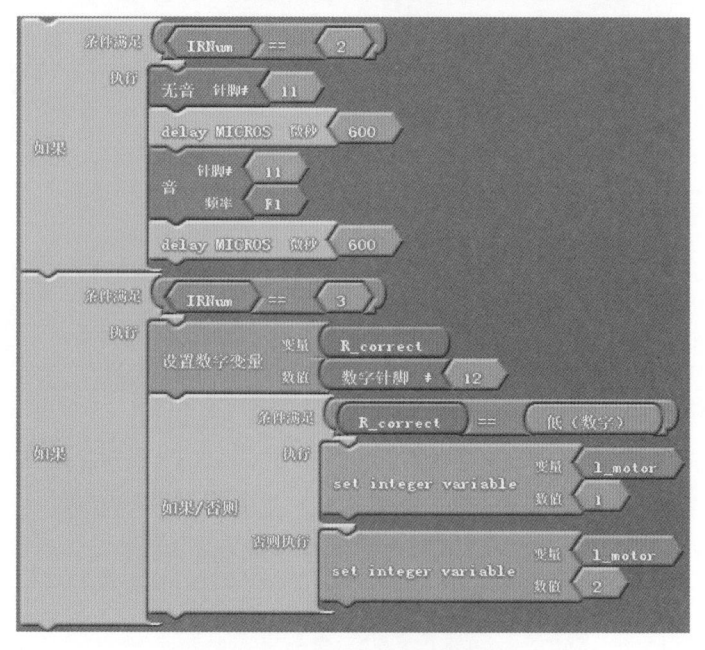

图3-2-20 右前传感器

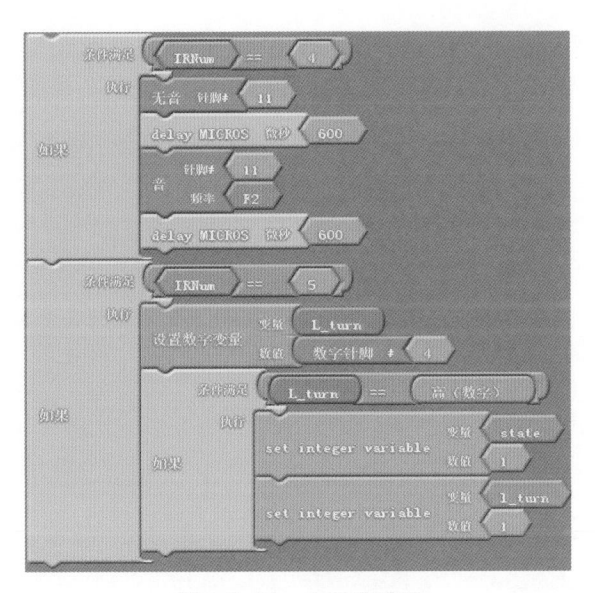

图3-2-21　左后传感器

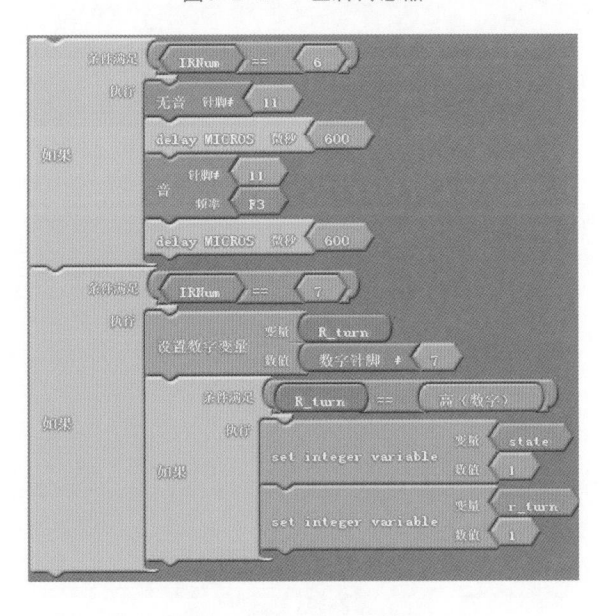

图3-2-22　右后传感器

（3）添加多线程程序（包含两个状态），如图3-2-23所示。

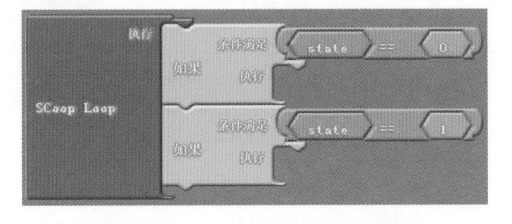

图3-2-23　两个状态的多线程

在这里两个"如果"图形也可以用一个"如果/否则"替换，但为了以后程序的功能优化，建议保留使用两个"如果"图形。

（4）校正走直线状态电动机控制，如图3-2-24所示。

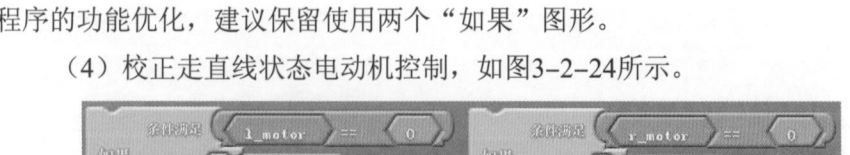

图3-2-24 校正走直线电动机控制

图3-2-25中左、右电动机的高速数值（150，165）是无校正走直线时测得的数据，低速数值（120，143）是在"智能鼠跑起来"任务中测得的数据。

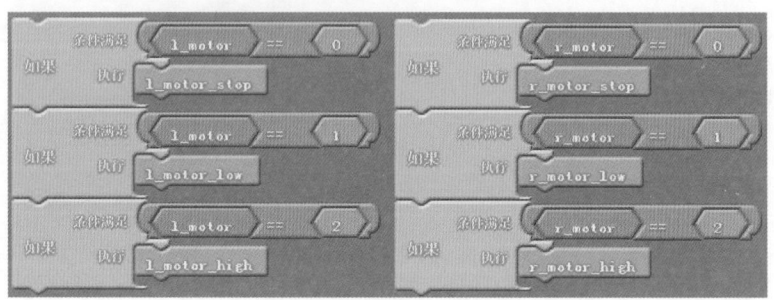

图3-2-25 左、右电动机的低速和高速

左、右电动机停止程序如图3-2-26所示。

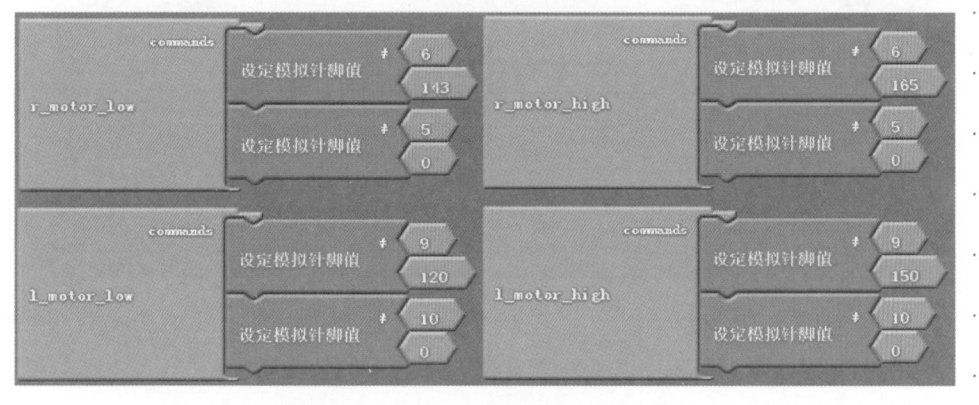

图3-2-26 左、右电动机停止程序

（5）转弯状态电动机控制。为了使转弯过程清晰地展示出来，在转弯前和转弯后分别添加0.2 s的停止程序（见图3-2-27）。

由于智能鼠采用的是原地转弯，当智能鼠转弯结束后，其后方传感器仍然处于路口中，所以添加0.35 s的强制直行程序，使智能鼠走过当前路口，避免

因检测失误而出现原地转圈的情况（见图3-2-28）。

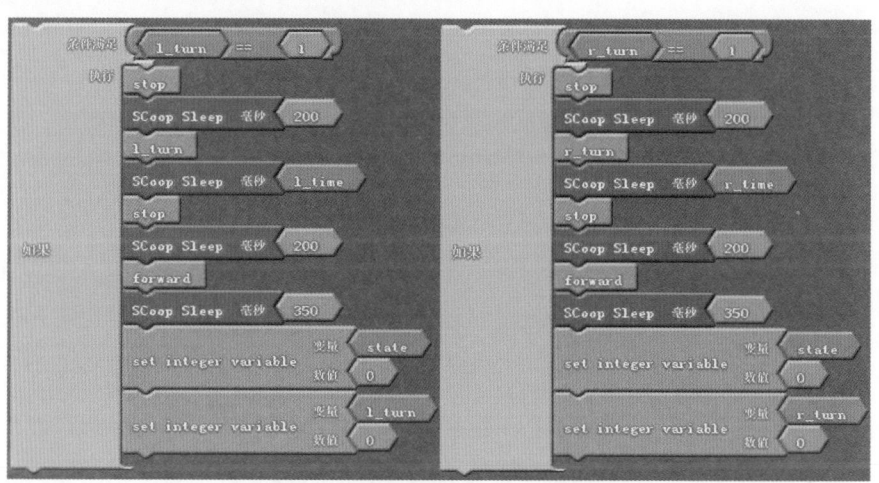

图3-2-27　转弯状态电动机控制

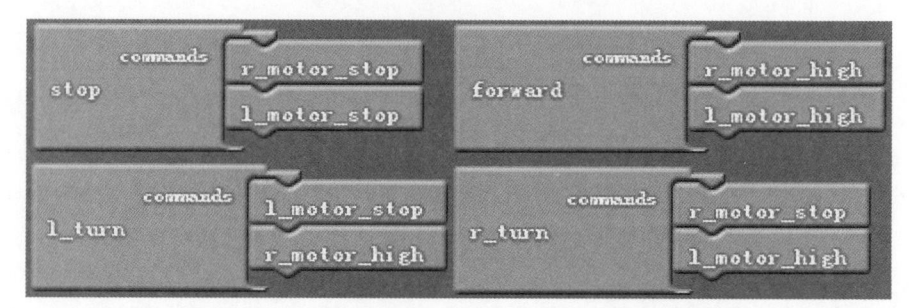

图3-2-28　智能鼠停止、前进、左转与右转子程序

（6）组合并下载程序。观察智能鼠在迷宫中的运行情况。

思考与总结

（1）针对智能鼠原地转弯姿态不够稳定的情况，应该如何优化？

（2）传感器的交替发射使得智能鼠在有限硬件资源的情况下，可以获得足够多的检测信息，从而提高运行效果。多线程技术的应用，使智能鼠传感器检测与电动机控制分离为两个独立的系统，避免相互之间的干扰，使系统响应速度大大提升。

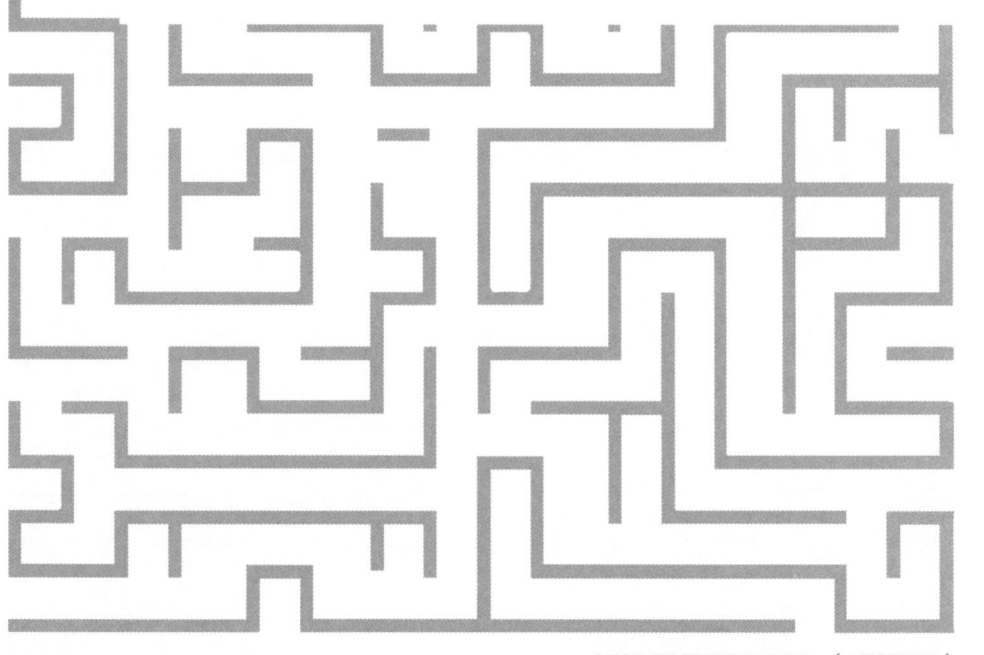

第四篇　拓展应用篇

　　本篇介绍了通过运用智能鼠核心技术，培养学生的自主探究和动手解决问题的能力。掌握如何在基础技术上灵活搭配，组合扩展功能。培养学生的设计、实践能力，锻炼学生的"创新、创意、创造"精神。

项目一

TQD-IOT工程创新课程平台结构组成

 学习目标

（1）了解创新实训平台和智能鼠的关系。

（2）学习TQD-IOT工程创新课程平台的硬件组成。

（3）学会使用TQD-IOT工程创新课程平台进行典型的物联网应用开发。

智能鼠涉及多种技术，可以很方便地扩展应用于其他场景。本项目主要介绍以智能鼠所包含的技术为基础进行扩展，延伸而成的TQD-IOT工程创新课程平台。

任务一 工程创新课程平台和智能鼠的关系

通过前三篇的基础知识学习，读者已经掌握了Arduino软硬件的使用和控制，以及控制器、蓝牙、红外传感器和电动机的使用方法。TQD-IOT工程创新课程平台，如图4-1-1所示，使用与智能鼠相同的控制器，在执行器和传感器方面进行了大量补充，表4-1-1所示为两者的对比。

图4-1-1 TQD-IOT工程
创新课程平台

表4-1-1 TQD-IOT工程创新课程平台与智能鼠竞赛平台对比

对比项目	TQD-IOT工程创新课程平台	智能鼠竞赛平台	对比结果
开发环境	Arduino开发软件	Arduino开发软件	相同
控制器	ATmega328P芯片	ATmega328P芯片	相同
执行器	直流电动机、舵机	直流电动机、步进电动机、减速电动机、伺服电动机等	递增
传感器	红外传感器、超声波传感器、振动传感器、光敏传感器、LED、LCD、蓝牙模块等30余种	红外传感器	递减
智能算法	无	左手法则、右手法则、中心法则等优化算法	不同

通过对比，可以看出TQD-IOT工程创新课程平台和智能鼠有相同点又有不同点，是智能鼠知识的延续和扩展，对于学习电子系统设计和应用具有相辅相成、递进补充的作用。

TQD-IOT工程创新课程平台特点如下：

1. 系统采用国际开源Arduino软硬件平台

CPU采用高性能、低功耗的AVR ATmega328P微控制器，内置蓝牙模块，最大支持6路模拟信号输出和14路数字信号输出（兼容PWM输出）。

2. 国际通用趣味图形化编程方式，易学易懂

创新实训平台软件支持代码和图形化两种编程方式，其中图形化编程可以实时转换为代码，从而有助力学生学习C语言等编程模式。图形化编程，模块种类变化多端，图形组合方式采用了彩色模块，提高了编程的趣味化。

3. 多媒体App在线学习资源丰富，形式多样

创新实训平台配套多媒体App，该App为自有知识产权软件，软件包括实验软件和在线学习资源两部分。实验软件针对每个章节设定为实验目的、软件仿真、实验器材、硬件连接、任务编程、效果演示、课后思考。在线学习资源通过现代智能远程学习方式，让学生做到学而知其用，用而知其在，在而知其所，所而知其代，代而知其原，原而知其衍。涵盖所有教学资源（各章节的详细讲解、程序、图片、视频），让学生随时随地自由学习。采用立体化教学手段，最大限度调动学生的学习积极性。

4. IOT模块化设计，扩展功能强大

创新实训平台采用模块化，分为控制、显示、操作三大区域。可选IOT模块包含：AVR控制模块、蓝牙通信模块、单色及多色LED模块、按键模块、点阵模块、超声波模块、有源及无源蜂鸣器模块、温度传感器模块、光敏传感器模块、风扇电动机模块、红外对射模块、语音识别模块等，涵盖典型物联网核心传感器和执行器应用，轻松实现各种智能家居的模拟智能控制。

5. 模块化磁吸附结构设计安全可靠、方便

创新实训平台各模块采用磁吸附式结构设计，多触点磁铁弹簧针接触紧密，无须连接导线，吸附力强，接触牢固，安全便捷。电连接采用20引脚弹簧针和凹槽碰触设计，创新实训平台连接图如图4-1-2所示。

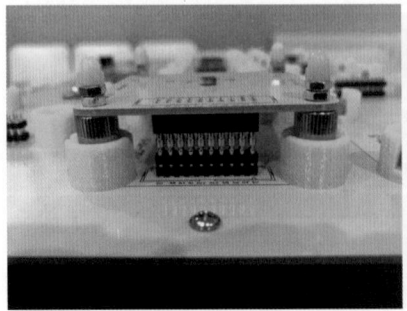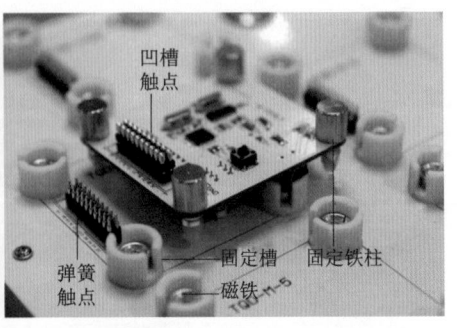

图4-1-2 创新实训平台连接图

平台2×4操作区同时支持8个IOT通用模块联动实验，该结构方便学生插拔，提高了连接的可靠性，延长产品使用寿命。

6. 无线蓝牙通信，方便快捷有效

控制模块集成了无线蓝牙模块，通信距离最远可达10 m，配套有独立的手机蓝牙App，方便用户与设备之间进行数据传输。

任务二 工程创新课程平台的硬件组成

创新实训平台分为三部分，即控制区、显示区和执行区。每个区域均可放置对应模块。模块与平台采用荣获国家实用新型专利的磁吸附设计，即放即通，无须导线连接，实验过程更方便，多达九个可操作位，轻松实现各种DIY实验设计。

一、控制区

用于放置控制器模块。核心控制器采用了与智能鼠相同的技术，如图4-1-3所示。该控制含有数字输入/输出接口和模拟输入/输出接口，通过其上一个20引脚端口（又称天线顶针）可方便地与不同的传感器、执行器和驱动器连接。

元器件都集成在控制器模块上，方便管理和操作，使用者可以通过蓝牙将实验平台与用户手机联系起来，实验过程当中可以在远端直接使用手机控制实验平台，节约成本和占地空间。

（1）核心控制器集成有蓝牙模块，用户可通过手机上安装的蓝牙串口软件对实验步骤进行指令操作，也可接收模块反馈的状态信息，从而实现信息的无线传输。

（a）核心控制器正面　　　　　　　　（b）核心控制器反面

图4-1-3　核心控制器

（2）核心控制器背面安装有AVR主控芯片、晶振、多组电容元件和电阻元件，所需核心元器件全部集成在电路板背面，使用时无须进行二次连接等操作。同时安装有复位键，可以将程序复位、清空，方便继续进行新的实验。

（3）核心控制器正面装有一键式开关，通电使用时无须进行其他操作，一键按下即可控制模块的工作状态。紧挨边缘的USB接口可直接与计算机连接，无须外接其他的下载器，将程序写入主控芯片当中来控制实验箱。

Arduino硬件平台基于AVR，对AVR库进行了二次编译封装，寄存器、地址指针之类均已经设置完毕，在使用时不需要进行特别设置。

Arduino硬件平台具有以下主要特点：

处理器：Atmel Atmega 328P。

数字 I/O：数字输入/输出端口D0～D13。

模拟 I/O：模拟输入/输出端口A0～A5。

支持ICSP下载，支持TX/RX。

输入电压：USB接口供电或者5～12 V外部电源供电。

输出电压：支持DC 3.3 V和5 V输出。

其中，数字接口的D3、D5、D6、D9、D10、D11（见图4-1-4）可以兼作PWM输出接口。

除D0、D1用于串口通信外，控制模块的其他所有引脚全部引出，方便自由组合进行实验。

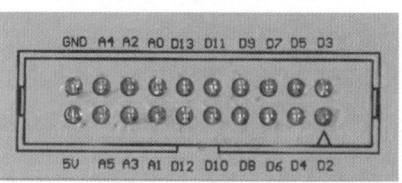

图4-1-4　数字接口

在日常生活中有很多设备，如笔记本式计算机、音箱、耳机等，都内置

了蓝牙模块，以便能够实现无线传输的功能。在实验里采用的蓝牙模块型号为HC-06（见图1-2-5）。通信距离最远可达10 m，极大方便了用户与设备之间的数据无线传输。

二、显示区

日常使用的计算器、电子表、万用表等很多电子产品中，都包含液晶显示模块，它具有显示质量高、体积小、质量小、功耗低等特点。本平台采用1602液晶显示模块，如图4-1-5所示，最大支持2×16个字符的显示，清晰度可调节、延时低、实时性好，可以准确地显示各类数据或信息。

图4-1-5 液晶显示模块

液晶显示模块与核心控制器的引脚连接如表4-1-2所示。

表4-1-2 液晶显示模块与核心控制器的引脚连接

序号	液晶显示模块数据通信引脚	核心控制器
1	VCC	VCC
2	GND	GND
3	RS	D12
4	EN	D11
5	IO	D5
6	IO	D4
7	IO	D3
8	IO	D2

这些引脚是固定的，不可更改；在使用液晶显示模块时这些引脚就无法再使用。

三、执行区

本平台共提供了2×4个可操作模块区，如表4-1-3所示。可根据实验实际需要自由增减模块种类或数量，实现"创新、创意、创造"。所有模块的连接均采用磁吸附的方式，有效避免了设备的损坏。

表4-1-3 执行区可操作模块

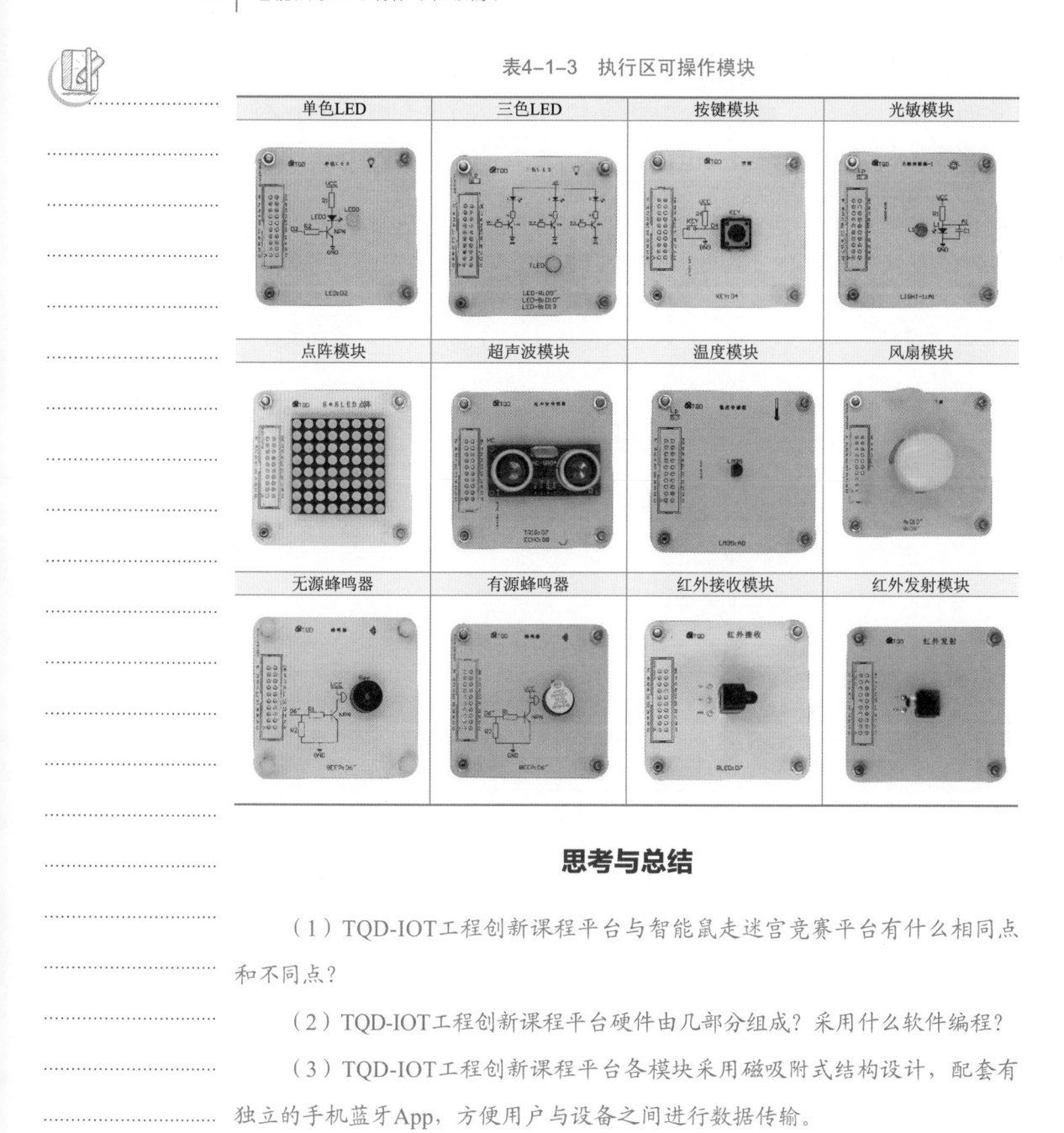

单色LED	三色LED	按键模块	光敏模块
点阵模块	超声波模块	温度模块	风扇模块
无源蜂鸣器	有源蜂鸣器	红外接收模块	红外发射模块

思考与总结

（1）TQD-IOT工程创新课程平台与智能鼠走迷宫竞赛平台有什么相同点和不同点？

（2）TQD-IOT工程创新课程平台硬件由几部分组成？采用什么软件编程？

（3）TQD-IOT工程创新课程平台各模块采用磁吸附式结构设计，配套有独立的手机蓝牙App，方便用户与设备之间进行数据传输。

项目二

智能鼠技术IOT扩展应用

学习目标

（1）学习灯光的智能控制以及多种信号的共同控制方法。

（2）学习常见安防的原理和实现方法。

（3）了解什么是视觉暂留。

（4）尝试DIY图像显示。

本项目使用TQD-IOT工程创新课程平台，通过"智能鼠控制器对灯光系统的控制"、"智能鼠控制器对安防系统的控制"和"智能鼠控制器对显示系统的控制"三个任务来模拟智能鼠包含的技术在现实生活当中是如何应用的。

任务一　智能鼠控制器对灯光系统的控制

近年来，随着照明建设逐步发展与成熟，智能控制系统也逐渐被引入照明管理系统。智能灯光系统是对灯光进行智能控制与管理的系统，与传统照明相比，它可实现灯光软启、调光、一键场景、一对一遥控及分区灯光全开全关等管理，并可用遥控、定时、集中、远程等多种控制方式，甚至用计算机来对灯光进行高级智能控制，从而达到智能照明的节能、环保、舒适、方便的功能。

子任务一　灯光的多彩变换

一、应用场景

随处可见的灯光，有单一颜色的，比如教室里照明用的荧光灯，报警的红灯，汽车上的黄色雾灯等；还有很多颜色不断变化的灯，如指示交通的红、黄、绿灯，如图4-2-1所示，商场中用作装饰的霓虹灯等。灯光是城市的美容师，每

图4-2-1　三色交通灯

当夜幕降临时，华灯初上，五颜六色的灯光就把城市装扮得格外美丽。那么，这些不同颜色灯是如何实现的呢？不同颜色之间又是如何切换的呢？

二、硬件介绍

1. LED简介

LED英文全称为light emitting diode，中文名称为发光二极管。它是一种能将电能转化为可见光的固态半导体器件。在日常生活中，LED的用途非常广泛。做实验时，常用的LED外形如图4-2-2所示。

LED有多种外形尺寸，它具有单向导电性，外加上合适的正向电压，就可以发光。根据制作材料或填充气体的不同，可以发出红色、绿色、蓝色、黄色等颜色的光。不同颜色的LED相互混合又可以形成其他的颜色，比如双色灯、三色灯甚至是全彩灯。

三基色混合原理如图4-2-3所示。

2. 模块介绍

三色LED模块：在实验中用到的三色小灯模块，供电端连接到一起，称为共阳极设计，如图4-2-4所示。

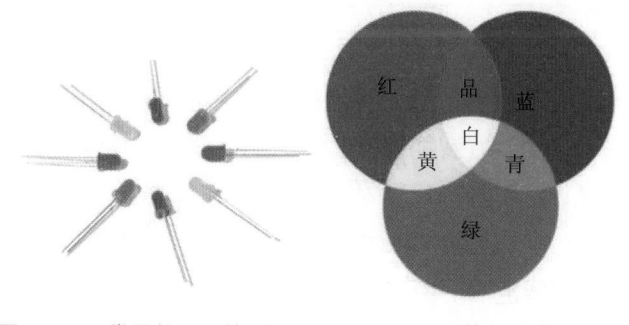
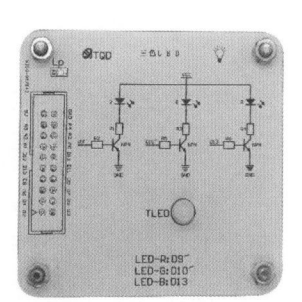

图4-2-2　常用的LED外形　　图4-2-3　三基色混合原理　　图4-2-4　三色LED模块

三色LED模块可以理解为红色、绿色、蓝色三个单色LED小灯放置在一个罩子里。当红色LED点亮时发出红光，绿色LED点亮时发出绿光，蓝色LED点亮时发出蓝光，当按照不同种类混合时，就可以发出其他颜色的光。三组单色小灯电路均可以单独控制，当输入高电平时导通，输入低电平时熄灭。

如表4-2-1所示，通过控制三色LED模块三个引脚的高低电平，可实现不同颜色的变换。

表4-2-1 三色LED模块引脚

序号	三色LED模块数据通信引脚	核心控制器
1	VCC	VCC
2	R	D9（Red，红基色）
3	G	D10（Green，绿基色）
4	B	D13（Blue，蓝基色）

三、操作步骤

下面，通过一个三色LED模块和按键模块介绍如何营造一个多彩的氛围。

1. 流程图

在开始编程之前，先思考一下整体流程：三色LED模块可以点亮不同的颜色，按照红色、绿色和蓝色的顺序变换颜色，如图4-2-5所示。

2. 图形化编程

（1）首先，程序的运行必须要从主函数开始，选择"控制"栏中的主程序模块，如图4-2-6所示。

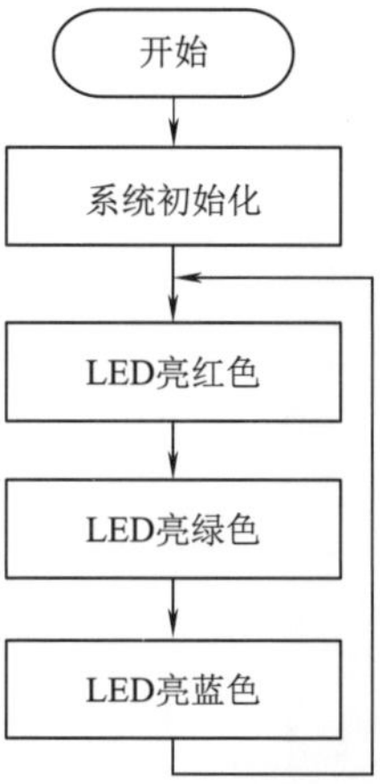

图4-2-5 营造色彩氛围流程图

图4-2-6 主程序模块

（2）设置三色LED模块的引脚电平，如图4-2-7所示。

图4-2-7 两种数字引脚图形

注意：第一个仅仅是读取状态，所以传感器使用这种图形；第二个是对引脚设置高低电平，所以执行器选用这种图形。

如图4-2-8所示，当IO 9输出高电平，IO 10、IO 13输出低电平时，亮红色。

当IO 10输出高电平，IO 9、IO 13输出低电平时，亮绿色。

当IO 13输出高电平，IO 9、IO 10输出低电平时，亮蓝色。

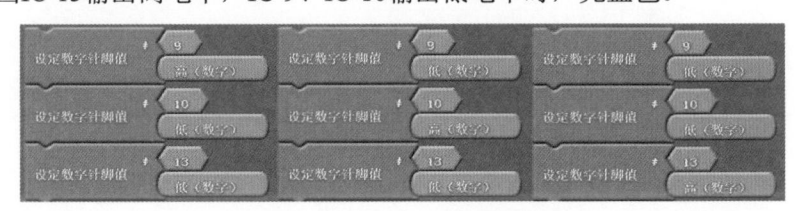

图4-2-8　红色、绿色和蓝色

（3）颜色变换。思考一下，直接将上面的程序组合起来可以实现三种颜色的变换吗？

如图4-2-9所示，下载程序后发现，三色LED亮出了白光。这是由于程序运行速度非常快，对于人眼来说几乎是同时接收到红色、绿色和蓝色，三种颜色等比例混合就成为了白色。

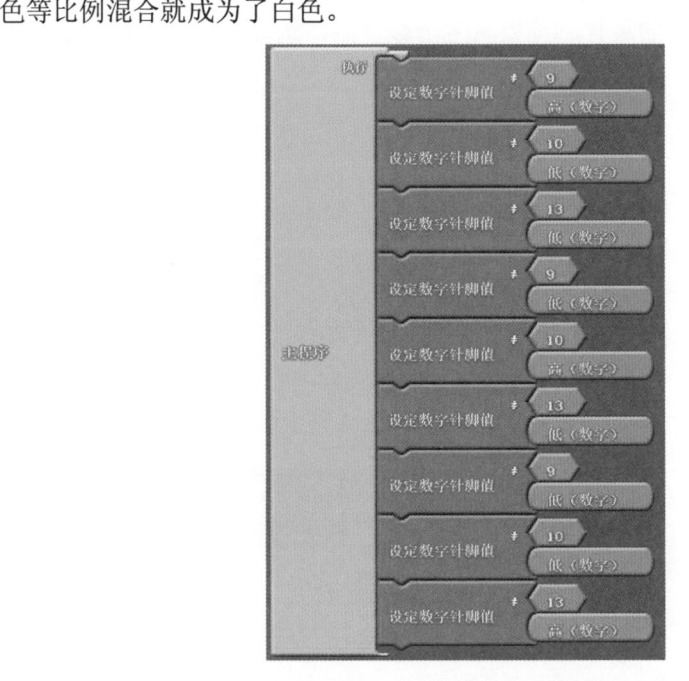

图4-2-9　直接组合三种颜色

所以，需要采用某种手段将颜色区分开来，使三种颜色发光时间长一些。

"控制"栏中的delay图形正好可以满足这一目的，如图4-2-10所示。

图4-2-10　delay

如图4-2-11所示，至此就完成了按键控制三色LED模块的颜色变换。

现在上传程序，观察一下效果。另外，读者可更改引脚9、引脚10、引脚13的高低电平，来观察三色LED的颜色变化。

注意：在上传程序时，控制模块上的白色按钮一定要按下去。

拓展训练：如何实现三色小灯亮出红、绿、蓝以外的颜色呢？（提示：可以用红、绿、蓝两两混合的方式实现。）

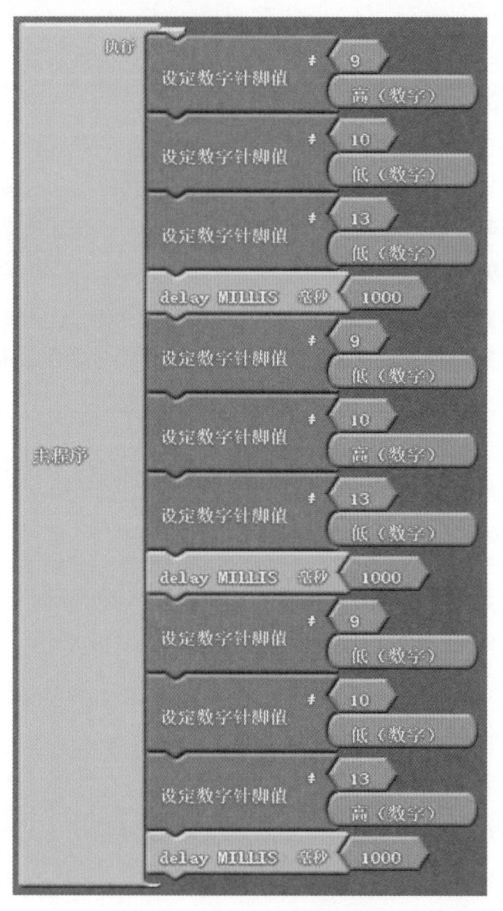

图4-2-11　三色变换程序

子任务二　手机遥控灯光

一、应用场景

随着日常生活节奏的加快，在日常生活当中，即时的操作已经无法满足人们对生活的追求。一种随时随地的可以远程操作的方法，成为人们渴望的目标。

随着物联网、云计算、无线通信等新技术的发展，智能家居得到了快速发展，更大地方便了人们的生活，如图4-2-12所示。用户利用智能手机来控制家中的设备，实现远程控制、场景控制、联动控制和定时控制等功能。上班时，可以远程监控家中情形，及时发现老人和小孩的状况；下班后，只需要远程发送指令就可以控制空调启动，电饭煲做饭，热水器开启，这样，当我们到家时，无须再花费等待时间，既可享受满满的生活。

图4-2-12　手机控制智能家居

二、硬件介绍

1. 无线通信原理

无线通信是利用电磁波信号在自由空间中传播的特性进行信息交换的一种通信方式。无线通信技术主要包括无线电通信、微波通信、红外通信和光通信等多种形式。其中，无线电通信的应用最为广泛，它是利用电磁波信号在自由空间传播的特性进行信息交换的一种通信方式。

蓝牙技术是一种常见的无线通信（数据）传输方式。具有低成本、低功率以及开放性强等特点。在汽车、手机以及计算机上都可以看到它的身影。

2. 模块介绍

蓝牙模块已经集成在控制模块上，避免了接线出现的错误。蓝牙模块在前文已详细介绍，这里不再赘述。

单色LED模块和三色LED模块非常相似，单色LED只可以发出一种颜色，如图4-2-13所示，依据填充气体的差异，可以发出不同的颜色。单色LED

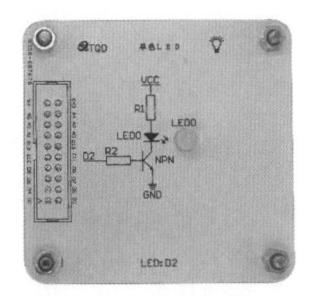

图4-2-13　单色LED模块

模块与核心控制器的引脚连接，如表4-2-2所示。

表4-2-2　单色LED模块与核心控制器的引脚连接

序号	单色LED模块数据通信引脚	核心控制器
1	VCC	VCC
2	GND	GND
3	IO	D2

三、操作步骤

下面，通过手机蓝牙APP来控制LED的亮灭，从而学习无线通信控制。

1. 流程图

在开始编程之前，先思考一下整体流程：首先手机调试工具（手机蓝牙App）要和集成在控制模块上的蓝牙进行连接（配对）；然后APP发送指令，控制模块读取这些指令，控制LED模块的亮、灭，如图4-2-14所示。

2. 图形化编程

（1）手机蓝牙App和控制模块蓝牙的连接，如图4-2-15所示。

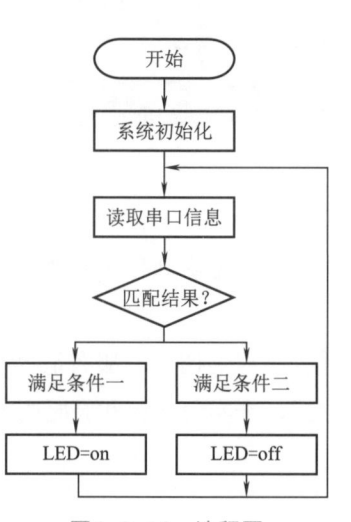

图4-2-14　流程图　　　　图4-2-15　手机App和蓝牙模块的连接

搜索新设备，并与控制模块蓝牙配对（配对密钥请查看标示贴，或尝试默认密钥"1234"）。

（2）蓝牙控制LED亮灭。前面已经介绍过智能鼠无线控制实验。下面依据学过的知识进行实验，如图4-2-16所示。

为了使实验过程更加清晰直观，修改手机蓝牙App发送的指令以及回传的

信息：

发送指令"o"，对应open，也就是打开LED，回传的信息修改为"LED is open!"

发送指令"c"，对应close，也就是关闭LED，回传的信息修改为"LED is close!"

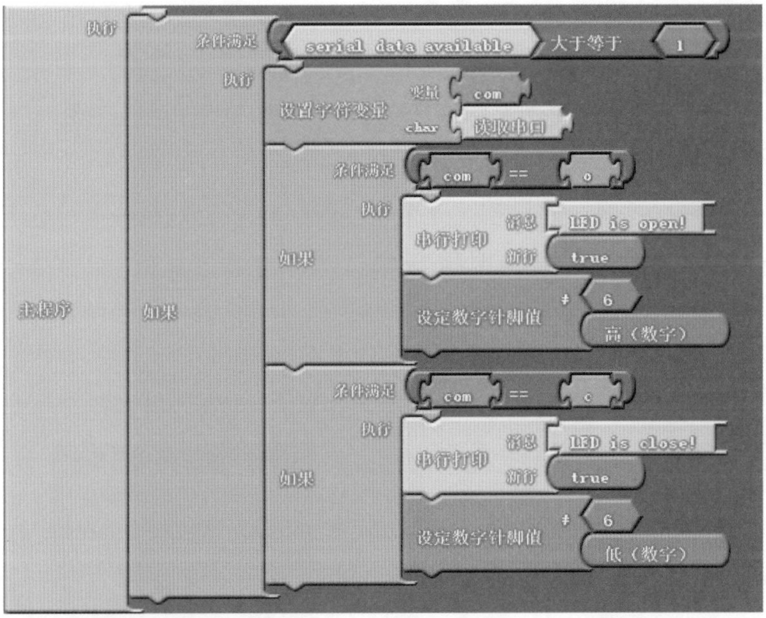

图4-2-16 蓝牙通信实验

注意：由于USB串口和蓝牙通信共用相同的RXD和TXD，所以在上传程序的时候一定要将控制模块上的白色按钮按下去，上传完毕进行蓝牙通信时，一定要将按钮弹起来。

拓展训练：上述实验已经实现了灯光的遥控亮灭的调节，那么是否可以添加新功能，实现灯光的遥控、自动两种调节方式，来实现亮灭和亮暗呢？（提示：采用光敏传感器。）

任务二 智能鼠控制器对安防系统的控制

随着经济发展，社会进步，人们的生活水平得到了很大的提高。享受生活之余，家居安全成为人们非常牵挂的事情。安防系统是实施安全防范控制的重要技术手段。在当前安防需求膨胀的形势下，其在安全技术防范领域的运用也越来越广泛。

随着光电信息技术、微电子技术、微计算机技术与视频图像处理技术等的发展，传统的安防系统也正由数字化、网络化，而逐步走向智能化。这种智能化是指在不需要人为干预的情况下，系统能自动实现对监控画面中的异常情况进行检测、识别，在有异常时能及时做出预/报警。

　　智能安防系统可以简单理解为：图像的传输和存储、数据的存储和处理准确而选择性操作的技术系统。一个完整的智能安防系统主要包括门禁、报警和监控三大部分。智能安防与传统安防的最大区别在于智能化。我国安防产业发展很快，也比较普及，但是传统安防对人的依赖性比较强，非常耗费人力，而智能安防能够通过机器实现智能判断，从而尽可能实现人想做的事。

　　近年校园安全问题深受各级教育、执法部门重视。加快校园安防系统的建设，提高防范、制止犯罪和应对突发事件的能力，是避免校园安全事故发生的有效手段。校园安防系统具备预防事故的功能，通过各种技术管理手段将校园安全隐患提前预判，有效避免校园安全事故的发生。以校园的实际情况进行分析，校园安防系统利用相关科技手段，使安全系统覆盖整个校园及周边区域，有效地帮助学校做好校园安全管理工作，如图4-2-17所示。

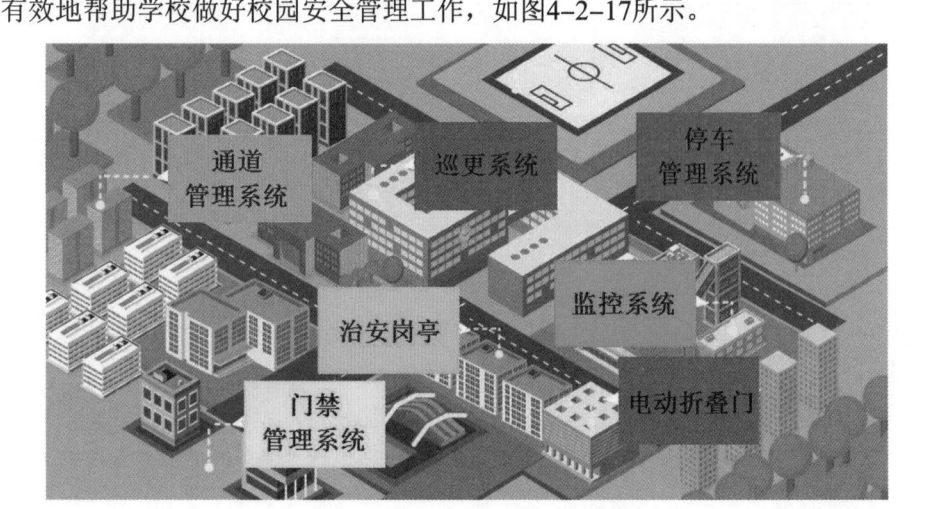

图4-2-17　IOT智控安防系统

子任务一　电子防护栏的应用

一、应用场景

　　在现实生活当中，很多情况都会用超声波作为判断条件，判断是否有物体靠近。

场景一：超声波传感器在运动检测中的应用。

超声波传感器近年在自动探测移动物体中得到更多的应用。超声波传感器主要利用多普勒效应进行工作，通过晶振向外发射超过人体能感知的高频超声波，一般典型的选用25~40 kHz波，然后控制模块检测反射回来波的频率，如果区域内有物体运动，反射波频率就会有轻微的波动，即多普勒效应，以此来判断照明区域的物体移动，从而达到控制开关的目的。

场景二：超声波传感器用于闸机系统的车辆检测。

在停车场和车库中通过安装传感器来检测车辆何时靠近挡板或车库中的其他障碍物，如图4-2-18所示。或者入口使用闸机系统来控制，当有车辆在栏杆下面，栏杆不能降下。超声波传感器特别适合控制这一过程。它们检测目标物不会受车辆的型号或者颜色的影响，在栏杆的下方监测整个区域。

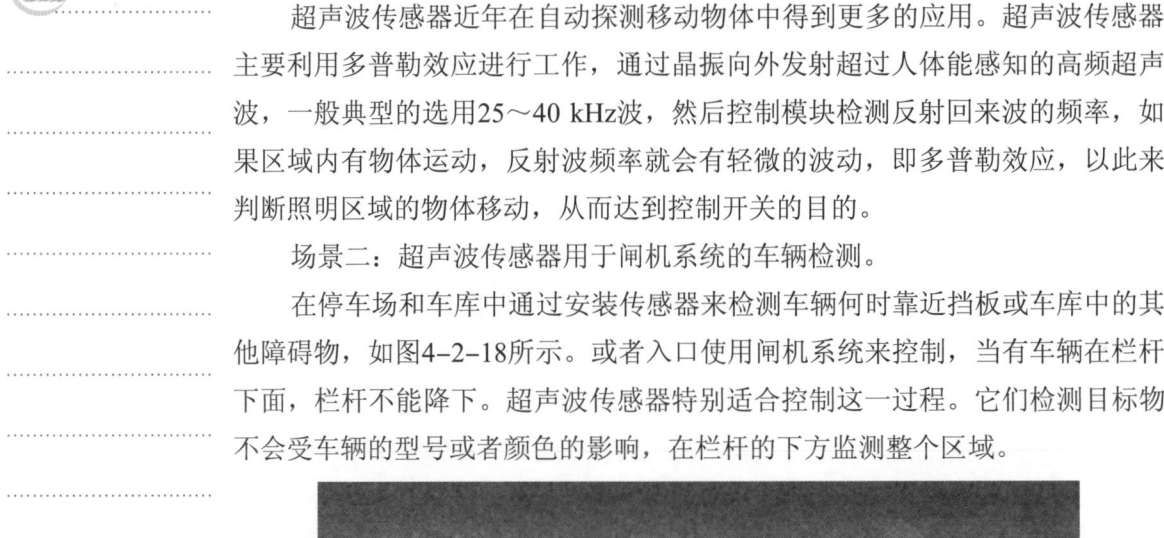

图4-2-18 超声波检测

二、硬件介绍

1. 声波简介

声波是声音的传播形式，它属于机械波。物体产生振动在空气或其他介质中的传播就称为声波。声波的传输必须依靠某种介质，在真空中是无法传输声波的。不同的介质，声波传输速度也不同。声波不只有速度，它还有频率，就是物体在单位时间内的振动次数。我们听到的声音有的悦耳，有的低沉，还有的尖锐刺耳。

各种声源发声的频率千差万别，使得声波丰富多彩。这些都是我们可以听得到的声音。自然界中还存在很多我们"听不见的声音"，如图4-2-19所示。

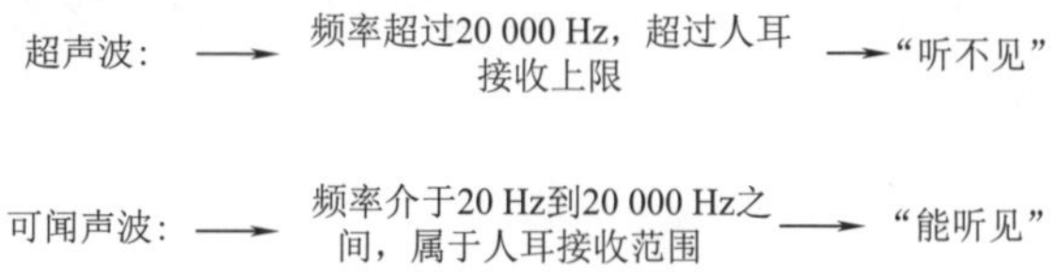

超声波：──→ 频率超过20 000 Hz，超过人耳 ──→ "听不见"
　　　　　　接收上限

可闻声波：──→ 频率介于20 Hz到20 000 Hz之 ──→ "能听见"
　　　　　　　间，属于人耳接收范围

图4-2-19　超声波和可闻声波的显著区别

根据声波频率的不同，可以分为以下几类：

频率低于20 Hz的声波称为次声波或超低声波。

频率为20 Hz～20 kHz的声波称为可闻声波。

频率为20 kHz～1 GHz的声波称为超声波。

频率大于1 GHz的声波称为特超声波或微波超声波。

2. 模块介绍

超声波传感器采用的是发射接收一体模块，如图4-2-20所示。

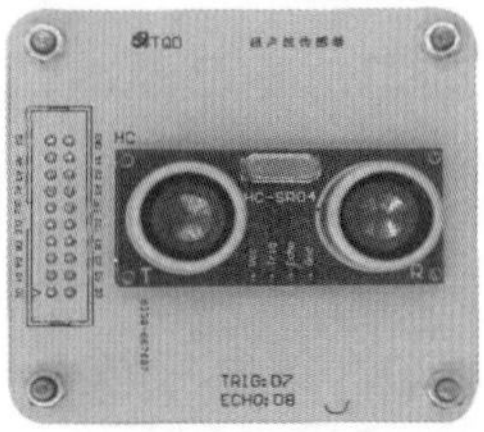

超声波传感器的引脚中VCC和GND是供电和接地端。

Trig是触发信号输入（有信号输入时，超声波发射）。

图4-2-20　发射接收一体模块

Echo是回声输出（超声波传感器接收到回声时，输出信号）。超声波模块与核心控制器的引脚连接如表4-2-3所示。

表4-2-3　超声波模块与核心控制器的引脚连接

序号	超声波模块数据通信引脚	核心控制器
1	VCC	VCC
2	GND	GND
3	Trig	D7
4	Echo	D8

超声波传感器输出的数值是以cm为单位。

前几个任务已经介绍过数字引脚和模拟引脚的区别。请猜测一下，超声波传感器需要使用模拟引脚还是数字引脚？

超声波测得的距离是在不断变化的，但它也仅仅是记录是否发出超声波，以及是否接收到超声波，再根据公式（时间差乘以声波速度）来计算距离。所以，超声波传感器使用的是数字引脚。

除此之外，在实验当中还会用三色LED模块，在色彩变换实验中已经讲解过，这里不再重复。

三、操作步骤

下面，通过电子防护栏实验来介绍超声波测距在智能安防中的应用。

1. 流程图

在开始编程之前，先思考一下整体流程：超声波模块作为检测装置，首先检测障碍物距离，然后对这个距离进行判断。假设大于15 cm时，是安全距离；小于15 cm时，进行报警，三色LED亮红色，如图4-2-21所示。

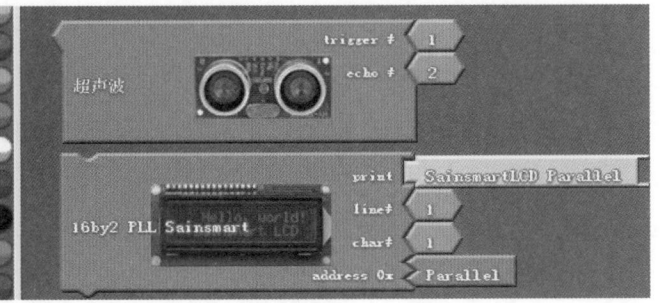

图4-2-21　报警流程图

2. 图形化编程

实验一：超声波测距实验。

（1）首先，在Generic Hardware栏中找到超声波模块和液晶模块。

注意： 实验平台上使用的是1602液晶模块，如图4-2-22所示。

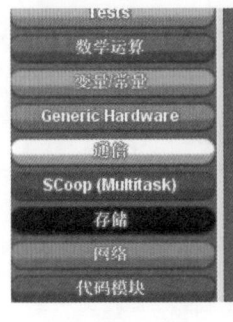

图4-2-22　1602液晶模块

前面已经学习过串行打印图形，对液晶图形的第一个引脚"print"是不是很熟悉？

它们都是将后面的信息显示出来。串行打印是通过串口监视器或蓝牙显示的；液晶是通过显示屏来显示的。

两个模块的通信同样需要使用glue模块来连接，修改超声波图形的引脚号，如图4-2-23所示。

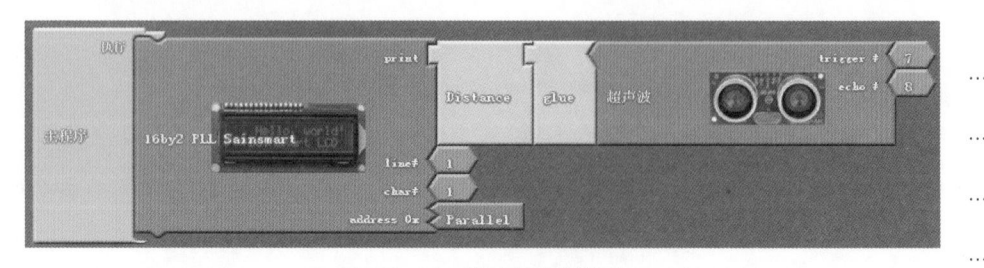

图4-2-23　超声波显示程序

上传程序，观察现场。

（2）如图4-2-24所示，当测距变化微弱时，数据显示影响较小；但是当测距变化较大时，非常不利于观察，并且数据出现10倍的偏差，这是什么原因？原因就是前一刻的数据和后一刻的数据几乎重叠到一起。

图4-2-24　测出的距离

例如，前一时刻的检测值是122 cm，下一时刻检测的实际距离是15 cm，但是液晶显示的数据会是152，这是因为液晶没有消除最后一位数字2，从而出现错误。

当变化频率过快时，数字还会重叠。由于人眼的视觉暂留特性，人们几乎看不清数据。所以，需要寻找一个解决的办法。

（3）首先需要对液晶进行擦除操作，也就是在显示数据之前都要擦除上一次的数据。

在Generic Hardware栏中正好有一个这样的模块，如图4-2-25所示。有多种功能可供选择。

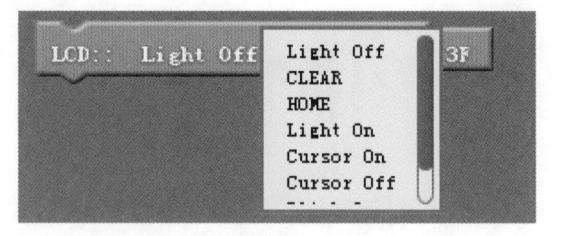

图4-2-25　擦除操作

选择CLEAR，并将通信方式修改为和1602液晶相同的Parallel，如图4-2-26所示。

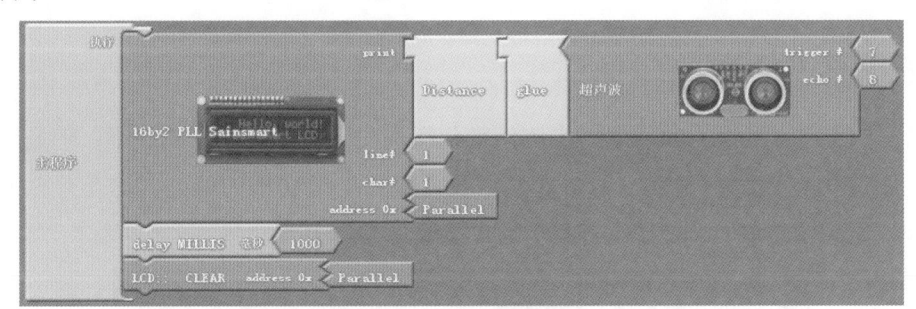

图4-2-26　修改通信方式

为了让电位器输出值刷新慢一些，添加了一个延时函数，如图4-2-27所示。

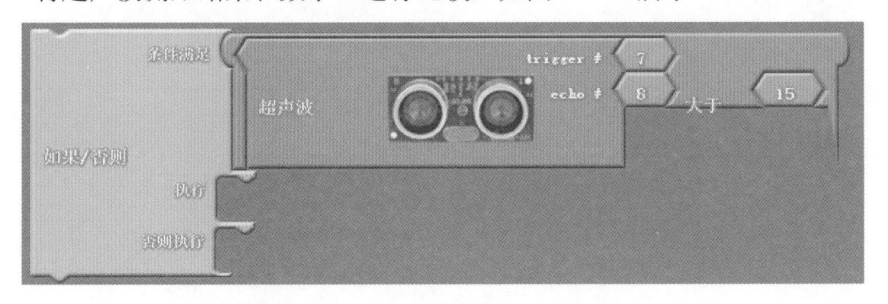

图4-2-27　主程序

注意：delay延时和CLEAR的前后顺序，如果CLEAR在delay前面，液晶会先执行擦除操作，这样在液晶上将几乎看不到数据。

实验二：电子防护栏实验。

现在已经知道超声波是如何测距的了，接下来就尝试对测距结果进行判断。

当距离大于15 cm时，三色LED亮绿色；当距离小于或等于15 cm时，三色LED亮红色。

这是一个条件判断，并且只有两种情况，所以选择"如果/否则"模块，如图4-2-28所示。

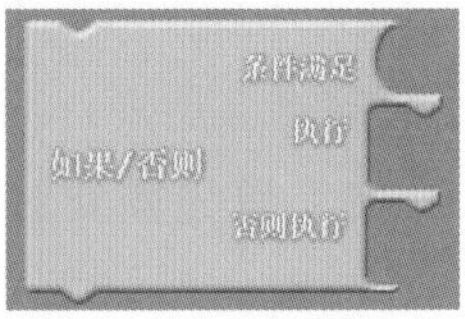

图4-2-28　"如果/否则"模块

将超声波测距结果和数字15进行比较，如图4-2-29所示。

图4-2-29　测距结果和数字15进行比较

将三色LED控制程序加入"如果/否则"模块当中，并连接主函数。

至此，电子防护栏实验就完成了，如图4-2-30所示。

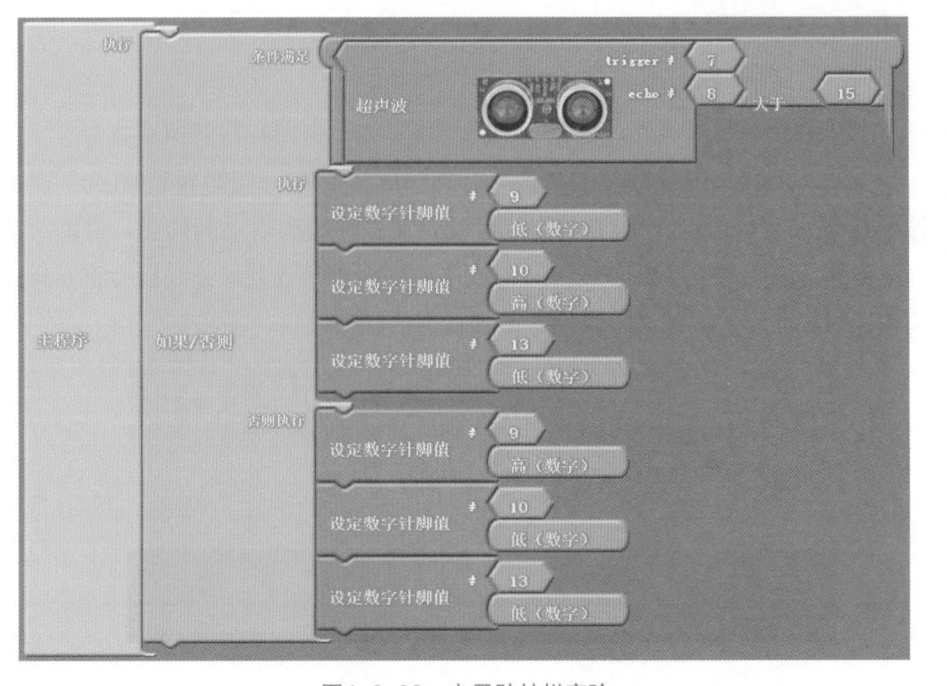

图4-2-30 电子防护栏实验

注意：下载程序的时候一定要按下白色按钮；液晶显示部分CLEAR和delay的前后顺序。

拓展训练：是否可以在程序当中添加无线通信呢？

当判断条件比较多时，比如10、15、20、30等多个判断值时，按照上面的编程方法还合适吗？如何精简？（提示：可以通过蓝牙向手机发送报警信息；将超声波进行变量赋值，这样，在后续的判断中，只需要调用变量就可以了。）

子任务二　门窗安防系统的应用

一、应用场景

随着科学技术的发展，人们的生活品质得到了极大提高，工作之余，拥有一个舒适的家是人们所重视的。人们的家庭防盗意识也越来越高。门窗是家居中必不可少的，为此，推出了一种门窗安防系统。

二、硬件介绍

门窗安防中应用较多的有一种称为红外线检测法。在门窗上安装红外对射

传感器，当有障碍物入侵时，对射被阻断从而发出警报。

1. 红外线简介

自然界中除可见光外，还有众多的不可见光，如图4-2-31所示。

可见光是指肉眼可见的光波域，从400 nm（紫光）到700 nm（红光），而波长760 nm到1 mm之间的光称为红外线，是一种肉眼看不到的光。借助一些光学设备，可以感受到红外线，通常红外线摄像机接收到红外线后会将其转化为可见的绿光。人们的肉眼永远见不到真正的红外线。

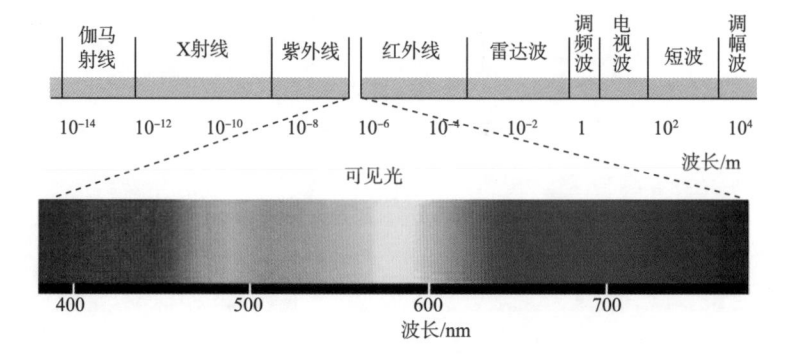

图4-2-31　光线类型

红外对射报警器安装于院墙上，当入侵者穿过红外对射传感器时，红外对射传感器随即发出报警信号。

2. 模块介绍

红外对射传感器由两部分组成，即红外发射头和红外接收头，如图4-2-32所示。透明的为红外发射头，黑色的为红外接收头。工作原理是根据红外接收头能否接收到红外发射头所发出的红外线，来输出信号，从而实现对其他电路的控制。红外发射头仅连接GND和VCC，起到红外线发射作用；红外接收头除要连接GND和VCC外，还需要连接信号线，如表4-2-4所示。

图4-2-32　红外对射传感器

表4-2-4　红外接收模块与核心控制器的引脚连接

序号	红外接收模块数据通信引脚	核心控制器
1	VCC	VCC
2	GND	GND
3	OUT	D7

当红外发射头所发出的红外线能够被红外接收头接收时，信号线输出低电平；否则，输出高电平。

蜂鸣器是一种一体化结构的电子讯响器，如图4-2-33所示。采用直流电压供电，广泛应用于计算机、打印机、复印机、报警器、电子玩具、汽车电子设备、电话机、定时器等电子产品中作为发声器件。

蜂鸣器分为有源蜂鸣器和无源蜂鸣器两种。这里的"源"不是指电源，而是振荡源，也就是说有源蜂鸣器内部自带振荡源，所以一通电就会

图4-2-33　蜂鸣器

鸣叫，使用相对简单；而无源蜂鸣器内部没有振荡源，必须使用PWM方波来驱动，使用不同的模拟量甚至可以让它发出不同的音调。

本实验是采用蜂鸣器作为报警设备，所以选用有源蜂鸣器即可，蜂鸣器模块与核心控制器的引脚连接如表4-2-5所示。

表4-2-5　蜂鸣器模块与核心控制器的引脚连接

序号	蜂鸣器模块数据通信引脚	核心控制器
1	VCC	VCC
2	GND	GND
3	IN	D6

三、操作步骤

下面，通过门窗安防系统实验来学习红外对射传感器在智能安防中的应用。

1. 流程图

在开始编程之前，先思考一下整体流程：类似超声波测距实验，当红外传感器完成对射时，输出高电平；被阻断时，输出低电平，以此作为判断条件，控制蜂鸣器的动作，如图4-2-34所示。

当红外线传感器完成对射时，蜂鸣器不发声，手机蓝牙App接收到信息safe。

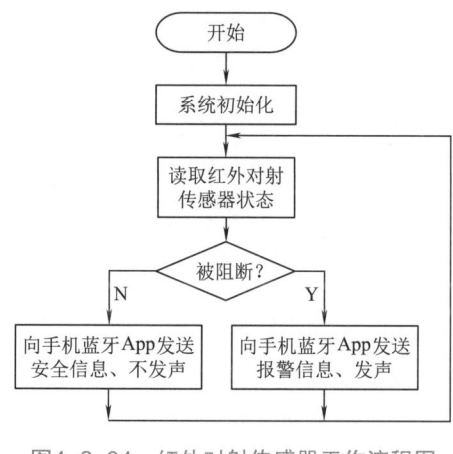

图4-2-34　红外对射传感器工作流程图

当红外线传感器被阻断时，蜂鸣器发声，手机蓝牙App接收到信息 warning！

2. 图形化编程

实验一：红外对射传感器输出实验。

下面通过一个小实验，来验证一下红外对射的输出电平和前面介绍的是否相同。

选择串行打印图形，通过串口来显示红外对射传感器的电平变化，如图4-2-35所示。（将message修改为level。）

如图4-2-36所示，经过验证，红外对射传感器完成对射时，输出0，也就是低电平；被阻隔时，输出1，也就是高电平。

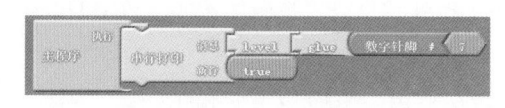

图4-2-35　红外对射输出

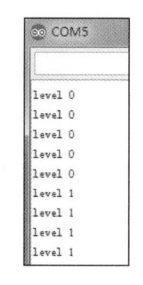

图4-2-36　输出参数

实验二：红外对射报警实验。

（1）由于红外对射传感器只有高电平和低电平两种输出状态，所以仍然使用"控制"栏中的"如果/否则"图形，添加判断条件，如图4-2-37所示。

图4-2-37　添加判断条件

（2）添加蜂鸣器的控制程序，如图4-2-38所示。

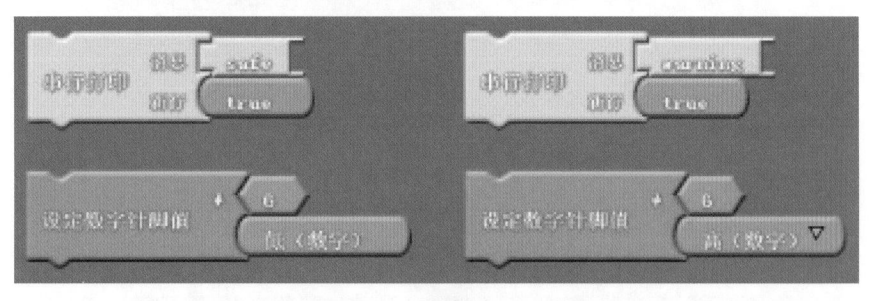

图4-2-38　两种情况安全和警报

（3）最后，将所有图形组合在一起，添加主程序图形，如图4-2-39所示。

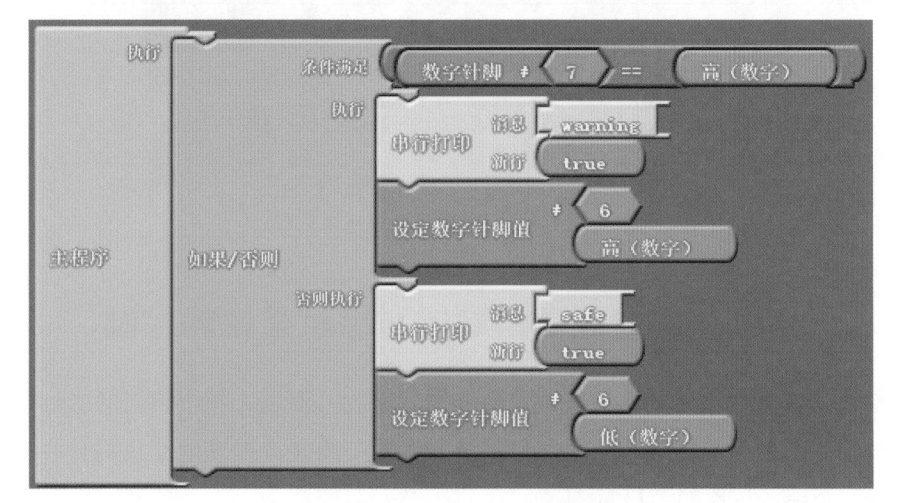

图4-2-39　添加主程序图形

注意： 串行打印图形在输出其他引脚数据时，记得使用glue图形进行连接；"如果"和"如果/否则"图形的区别；上传程序时一定要按下白色按钮，进行无线蓝牙通信时一定要弹起白色按钮。

拓展训练： 如何通过显示模块来输出safe或warning信息？如图4-2-40、图4-2-41所示。

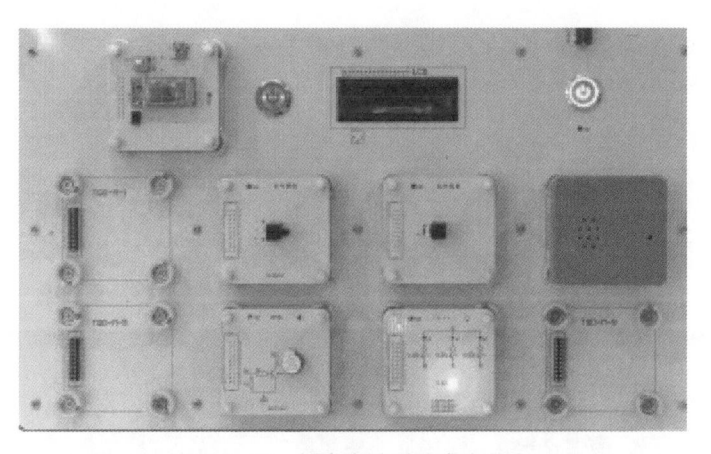

图4-2-40　门窗安防系统安全状态

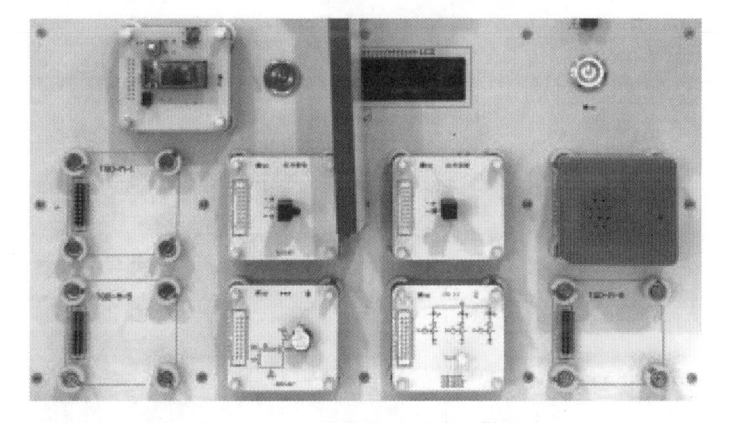

图4-2-41　门窗安防系统报警状态

任务三　智能鼠控制器对显示系统的控制

随着互联网技术的飞速发展，每当节日来临，商场、超市门口的彩灯装饰和炫彩的图像交相辉映、缤纷夺目，多元素的美感极大限度地让我们身心得到放松，营造了浓厚的节日氛围。

一、应用场景

在商场或是广场可以看到很多显示屏，有的显示文字，有的显示图案。近距离观察可以发现，这些显示屏都是由众多的LED小灯组成的，每一个LED小灯代表一个像素点，通过控制不同位置的LED亮灭不同的颜色，从而实现文字和图案的显示，如图4-2-42所示。

图4-2-42　LED点阵

　　LED显示屏常用的显示方法有逐列扫描（逐行扫描）和隔列扫描（隔行扫描）。逐列扫描（又称非交错扫描）是一种对位图图像进行编码的方法，通过扫描每列（行）像素，在电子显示屏上"绘制"视频图像。隔列扫描和逐列扫描类似，扫描方式是一列隔一列进行的，本次扫描奇数列，下一次扫描偶数列。就清晰度而言，逐列扫描更优，如图4-2-43所示。

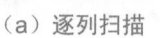

（a）逐列扫描　　　　　　　　　　　　（b）隔列扫描

图4-2-43　不同扫描方式

二、模块介绍

　　TQD点阵模块如图4-2-44所示，共阳极设计，高亮红色显示；由8×8共64个LED小灯组成。

　　按照传统的LED控制方法，整个8×8点阵需要64个I/O口。为了节省端口资源，对点阵模块进行了特殊设置，每行、每列均共用引脚IO，这样每个LED均对应不同的行和列，共计16引脚即可实现数据的显示。

　　当行引脚是低电平、列引脚是高电平时，对应LED才会点亮。

　　点阵模块数据通信引脚和控制器模块的对应关系如图4-2-45所示。

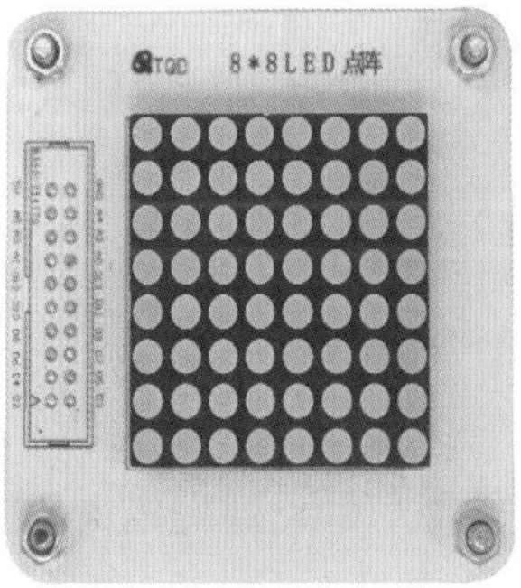

图4-2-44　点阵模块

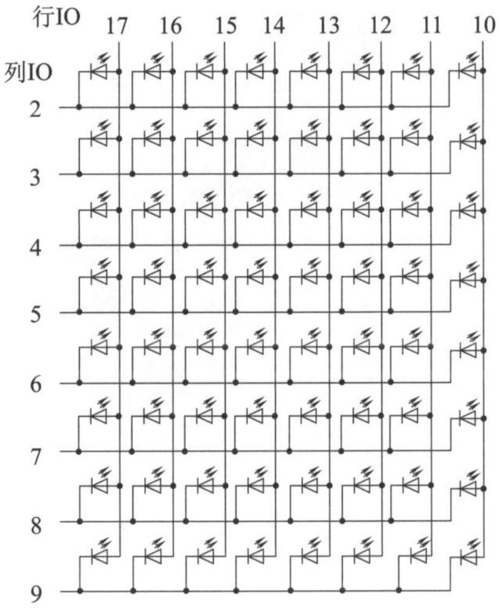

图4-2-45　点阵模块数据通信引脚图

其中，模拟端口A0～A3可以直接作为数字端口D14～D17使用。

三、操作步骤

下面，通过显示箭头滚动来学习屏幕显示的方法。

在开始编程之前，先思考一下整体流程：点阵模块的行和列共用I/O，选择较常见的逐列扫描进行实验。完成箭头静态显示（共八种），如图4-2-46所示，按照顺序排列即可实现箭头的滚动。

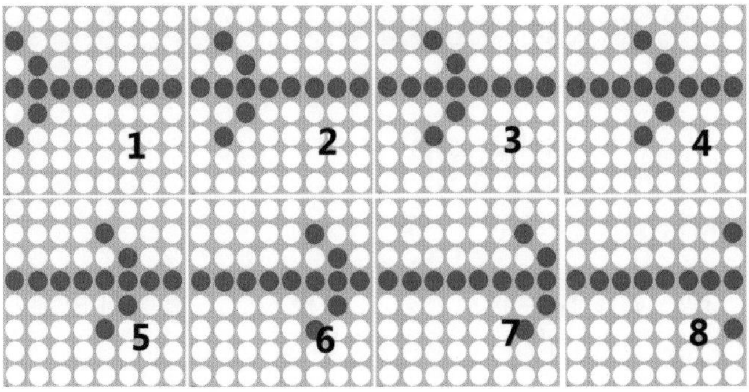

图4-2-46　箭头静态显示

1. 静态显示和动态显示流程图

静态显示和动态显示流程图，如图4-2-47所示。

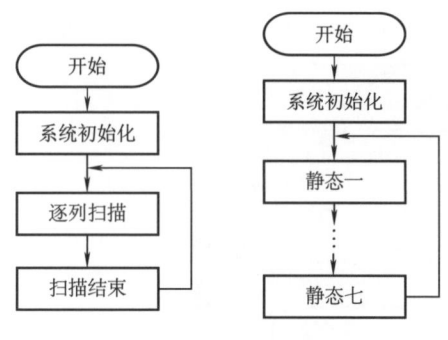

图4-2-47　静态显示和动态显示流程图

2. 图形化编程

下面通过一系列小实验来介绍如何使用点阵模块。

（1）点亮单个小灯。例如第4行（引脚5）第1列（引脚17），如图4-2-48所示。

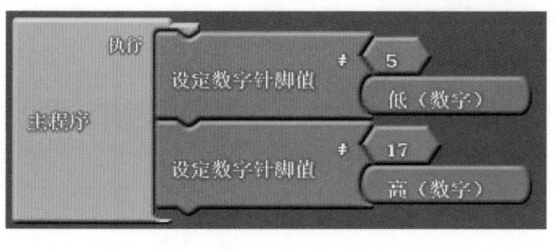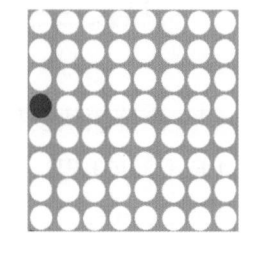

图4-2-48　点亮单个小灯

（2）点亮一行小灯。例如点亮第4行（引脚5）所有的LED小灯，如图4-2-49所示。

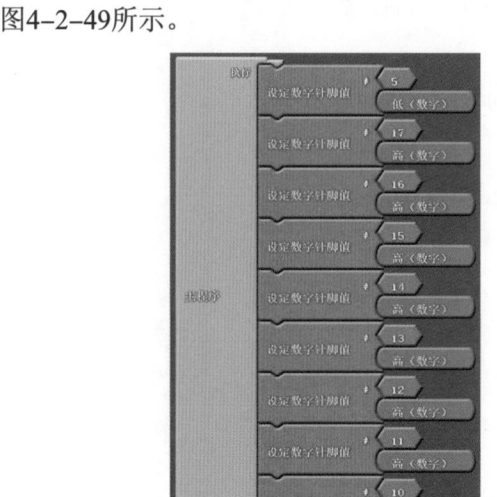

图4-2-49　点亮一行小灯

（3）点亮一列小灯。点亮第2列（引脚16）所有的LED灯，如图4-2-50所示。

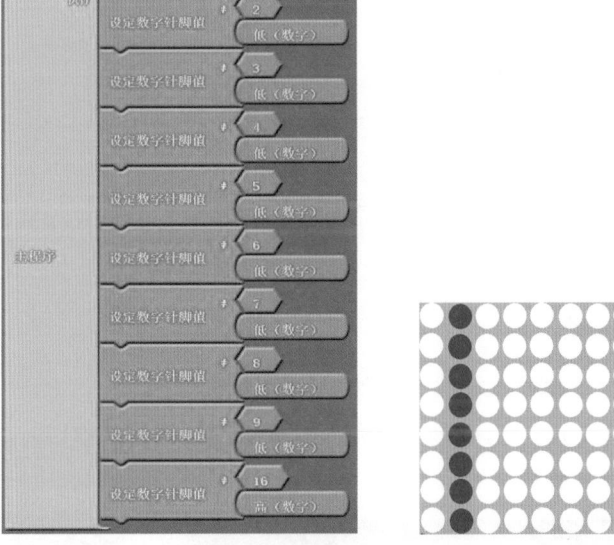

图4-2-50　点亮一列小灯

（4）点阵模块复位。在扫描的不同阶段，需要点亮不同的LED，为了避免图形的干扰（类似液晶的clean清屏功能），在交替之前需要对LED复位一次。

将64个小灯全部复位设置，采用电平反向设置的方法，即将所有的行设置为高电平、列设置为低电平，如图4-2-51所示。

现在读者已经学会点阵模块的使用了，下面尝试完成箭头的显示。

（5）箭头静态。首先来点亮图4-2-46中的第一个静态箭头，逐列扫描。

第1列，点亮第2、第4和第6个LED，如图4-2-52所示。

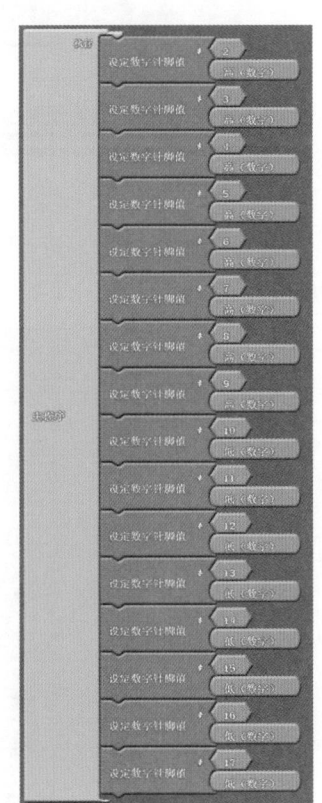

图4-2-51　点阵复位

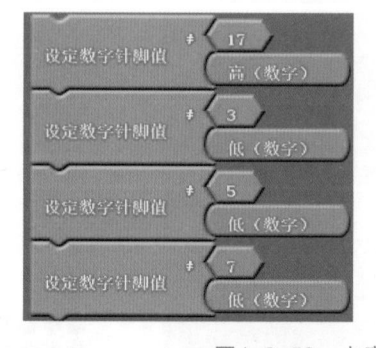
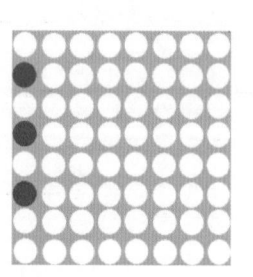

图4-2-52　点亮第1列

第2列，点亮第3~5个LED，如图4-2-53所示。

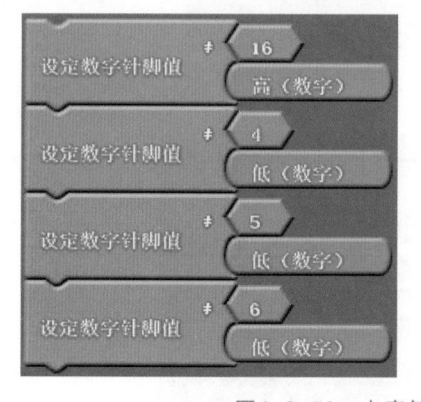
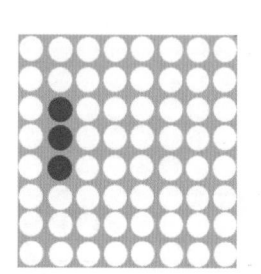

图4-2-53　点亮第2列

第3列，点亮第4个LED，如图4-2-54所示。

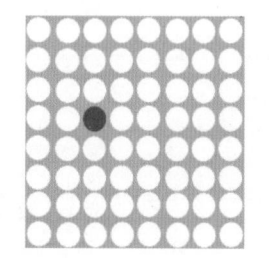

图4-2-54　点亮第3列

第4~8列，均点亮第4个LED，如图4-2-55所示。

在逐行扫描之间一定要添加复位，否则点阵将亮成一片，如图4-2-56所示。

将复位命名为子程序reset，那么第一张静态箭头程序如图4-2-57所示。

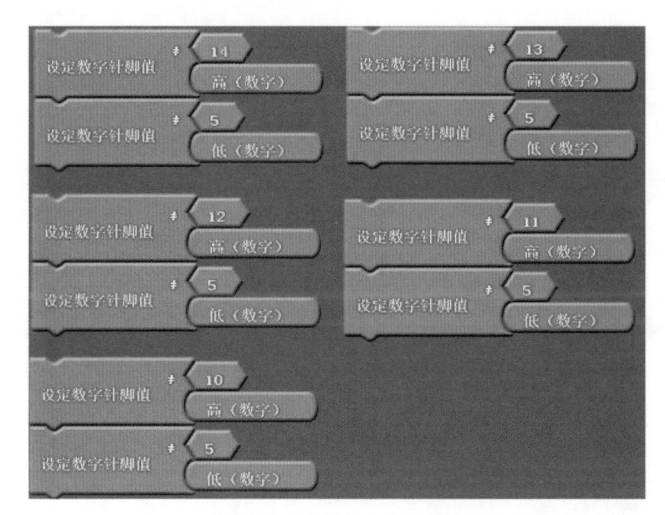

图4-2-55　点亮第4-8列

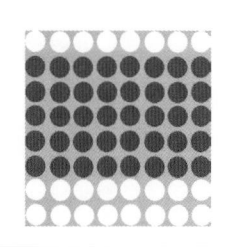

图4-2-56　未添加复位

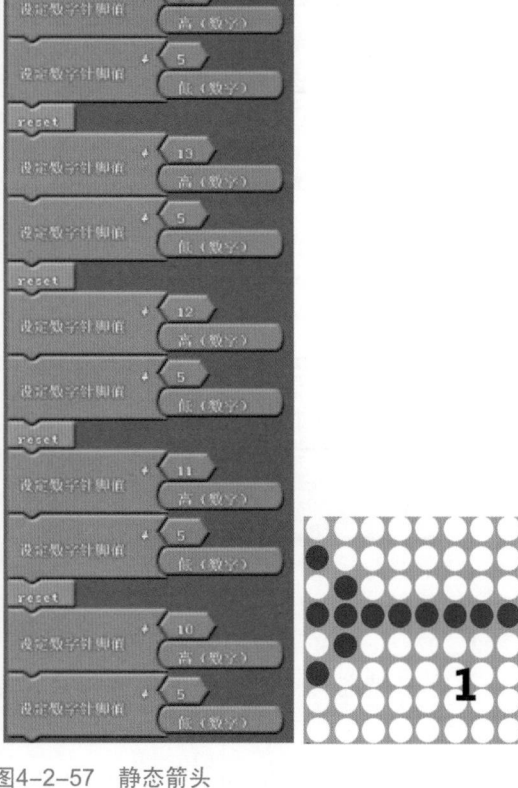

图4-2-57　静态箭头

对于其他七张静态箭头，读者可以自己通过修改8列显示的LED尝试一下。

（6）箭头滚动。完成上述八张静态箭头后，接下来就可以尝试让箭头滚动起来了。

将它们分别命名为子函数"one""two""three"……"seven""eight"。

① 现在来思考一下，如果直接将子函数添加到主程序当中，会是什么现象？

由实际现象可以看到，LED又亮成了一片，并没有显示出动态效果，如图4-2-58所示。

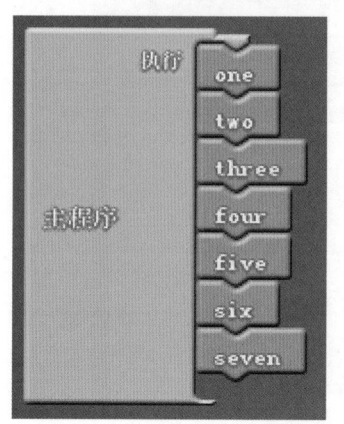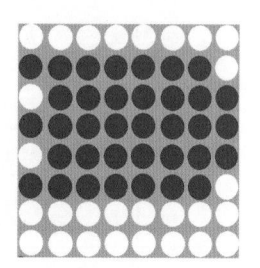

图4-2-58　静态箭头直接组合

这是什么原因呢？由于程序运行速度非常快，达到了毫秒甚至微秒级，人的肉眼分辨不出"one""two""three"……"seven""eight"八张静态箭头的交替运行。由于人眼的视觉暂留特性，人们的眼睛将这八张静态箭头叠加到了一起，就像单张静态箭头是由8列扫描图叠加而成的一样。

② 是否可以在子程序之间添加延时呢？静态箭头组合（50 ms延时）如图4-2-59所示。

选择延时50 ms，这样每秒可以传递20张图片，满足人眼对连续图像的要求（12张以上）。

上传程序观察现象，是不是只能看到箭头最后一列LED在发光？箭头其他部位的LED亮度非常低，几乎观察不到。

delay函数：保持控制器当前的输出状态不变，不进行其他操作，直到delay时间结束再进入下一个操作。

由于"one""two"……"seven""eight"子函数最后一部分都是第8列LED的扫描，箭头其他部位的LED在50 ms的时间内的占空比非常的低，亮度几乎不可察，所以图4-2-59的程序运行结果就像是第8列LED的发光。

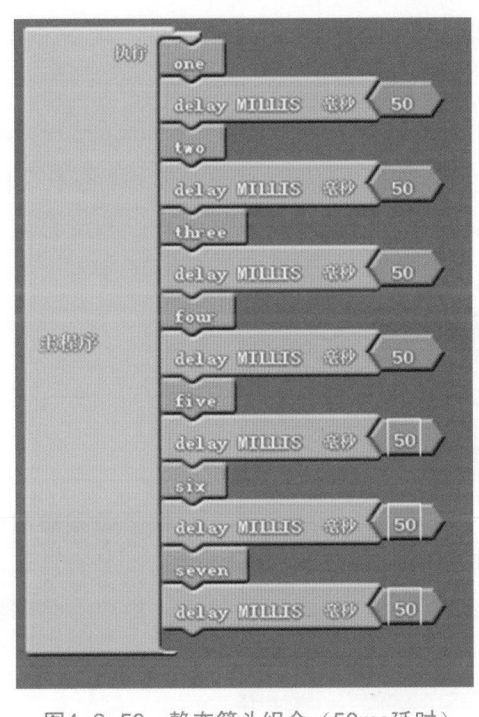

图4-2-59　静态箭头组合（50ms延时）

需要尝试其他方法来解决这一问题。

③ 最初添加delay函数的目的，是为了将八个静态箭头区分开，那可不可以通过分别多次运行"one""two"……"seven""eight"来实现呢？

"one""two"……"seven""eight"单次运行很快，在毫秒甚至微秒级别，那每张静态图重复运行100次、200次后再显示下一张图，这样是不是就可以肉眼可察了呢？

"控制"栏中共有三种循环图形，如图4-2-60所示。

图4-2-60　循环图形

第一种是先检查test是否为真，再去运行循环，否则直接跳过这一段程序。

第二种是先运行一次这段程序，再验证test是否为真，若为真则循环，否则进入下一段程序。

第三种是无条件循环，强制循环提前设定的次数，然后才会进入下一段程序。

在这里使用第三种无条件循环即可，设置次数为200，如图4-2-61所示.

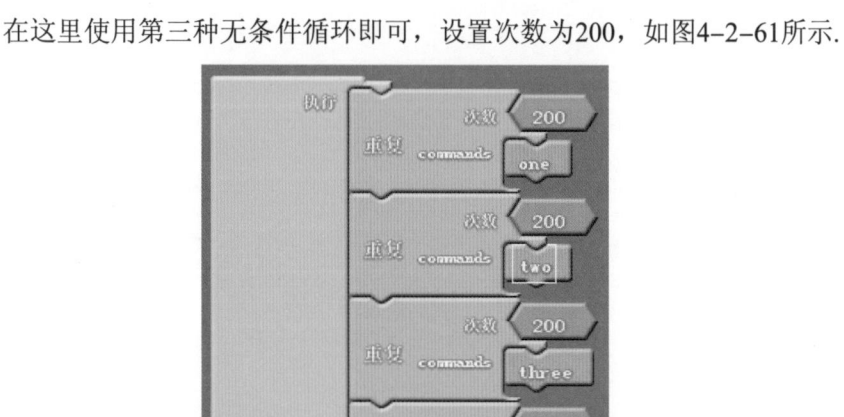

图4-2-61　逐个循环运行的箭头程序

上传程序观察结果，可以看到箭头滚动起来了。读者可以尝试修改循环次数来改变滚动的速度。

注意：箭头动态显示时，使用了人眼的视觉暂留特性。循环次数的大小决定了滚动的连贯性，可以多次尝试，直到找到一个合适的数值（以肉眼观察无闪烁为宜）。

拓展训练：是否有其他方法可以实现箭头的滚动呢？（提示：观察每列的LED引脚IO可以发现，列IO在逐个减小1，行IO只有3、5和7，可以尝试采用IO循环赋值的方法。）

读者还可以尝试心形实验。

思考与总结

（1）延时图形的作用是什么？在使用时应当注意什么？

（2）常见的无线通信方式有哪些？

（3）超声波检测和红外对射检测有什么区别？

（4）图像显示利用了人眼的视觉暂留特性，当每秒刷新的静态图像超过12张时，人眼就会认为图像是动态的。常用的显示方法可以分为行扫描和列扫描两种；逐行（逐列）相比较隔行（列）清晰度高，但软硬件要求也会更高。

附　录

TOD-MICROMOUSE-JO/JD/JM

附录A

风靡全球的国际智能鼠走迷宫竞赛

2019年是智能鼠走迷宫竞赛有史以来，最具兴盛发展、硕果累累的一年，在世界各地如火如荼地举行国际智能鼠走迷宫竞赛如图A-1所示。

1月，在印度孟买举办印度智能鼠国际竞赛。

3月，在美国加利福尼亚州举办APEC国际智能鼠竞赛。

4月，在葡萄牙Gondomar（贡多马尔）举办国际智能鼠走迷宫竞赛。

5月，在中国天津举办IEEE智能鼠走迷宫国际邀请赛。

6月，在英国伦敦举办智能鼠国际竞赛。

8月，在智利举办智能鼠走迷宫国际竞赛。

10月，在埃及举办埃及智能鼠国际竞赛。

11月，在日本东京举办全日本智能鼠国际公开赛。

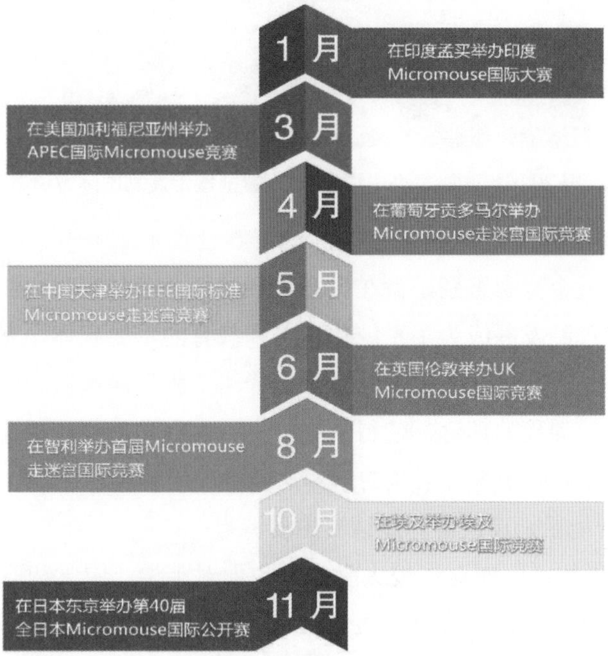

图A-1 国际智能鼠竞赛赛事安排

国际智能鼠走迷宫竞赛将成为全球高等教育、职业教育、普通教育，技术创新产教融合发展的助推器。在人工智能智能鼠走迷宫竞赛蓬勃发展的国际大环境下，教育领域适时地引进国际知名赛事提升学生的专业综合能力，掌握实践与创新的经验，助力产教融合发展，为行业、产业、企业培养更多优秀种子人才。

1. 中国IEEE智能鼠走迷宫国际邀请赛

从2009年开始，天津启诚伟业科技有限公司把智能鼠走迷宫竞赛引入中国，将IEEE智能鼠走迷宫竞赛进行本土化创新改革，对满足产业优化升级，开阔国际视野，掌握实践与创新经验，培育高技术高技能人才，起到了引领推动作用。

从2016年至2019年，连续四届举办中国IEEE智能鼠走迷宫国际邀请赛，该竞赛由天津市教育委员会主办，天津启诚伟业科技有限公司和天津渤海职业技术学院承办，如图A-2所示。

图A-2 从2016年开始举办"中国IEEE智能鼠走迷宫国际邀请赛"

目前，中国IEEE智能鼠走迷宫国际邀请赛设置了"中学、高职、本科、硕士、职业"共五个竞赛组别，旨在提升大赛的社会参与度和专业覆盖面。智能鼠走迷宫竞赛已经发展成为了系统化培养和教育的重要载体。充分体现光机电结合、软硬件结合、控制与机械结合，演绎"工程"课程概念的同时，延伸和扩展了"创新"课程的理念，使得学生的学习内容和教师的授课方式都有了全新的内涵，真正着眼于综合素质的培养，创造快乐素质教育。

中国IEEE智能鼠走迷宫国际邀请赛主要特点：

（1）参赛群体：既面对在校大学生，也面对小学、中学和职业人士，体现贯通式培养，终身教育的特点。还包括国际智能鼠专业级选手和历届国际智能鼠竞赛获奖选手。

（2）迷宫场地：既有面向中小学的8×8智能鼠迷宫场地，也有面向大专

● 视 频

中国IEEE智能鼠走迷宫国际邀请赛

院校的16×16全迷宫古典智能鼠场地；更有面对精英选手的25×32半尺寸智能鼠迷宫场地。体现竞赛的延展性，以智能鼠走迷宫竞赛为核心形式，不同学习阶段的学生都可以参赛。

（3）竞赛项目：既有智能鼠走迷宫赛项，又有自走车赛项，体现了竞赛既有技术性也有工程性，以工程应用为导向的竞赛思想。

（4）竞赛规则：普通教育、职业教育、高等教育、职业精英竞赛规则的相同点和差异点对比见表A-1。

表A-1 竞赛规则相同点和差异点对比

参赛类别	普通教育组	职业教育组	高等教育组	职业精英组
竞赛形式	（1）程序参数App在线调试。 （2）图形化趣味编程。 （3）IOT智能传感技术应用。 （4）8×8迷宫竞速	（1）理论知识考核。 （2）根据裁判现场任务编程并实现相应功能。 （3）现场技术答辩。 （4）16×16古典迷宫竞速	（1）DIY外观及结构机械设计。 （2）硬件技术创新。 （3）程序算法创新。 （4）16×16古典迷宫竞速	（1）DIY外观及结构机械设计。 （2）硬件技术创新。 （3）程序算法创新。 （4）25×32半尺寸迷宫竞速
竞赛内容	（1）组装任务10%。 （2）调试任务40%。 （3）竞速任务50%	（1）理论考核20%。 （2）创新赛30%。 （3）竞速赛50%	（1）创新赛20%。 （2）竞速赛80%	竞速赛100%

智能鼠国际专家现场培训指导如图A-3所示。

图A-3 智能鼠国际专家现场培训指导

2. 美国APEC世界智能鼠竞赛

1977年在美国纽约举行的首场令人震撼的智能鼠走迷宫竞赛，由IEEE与APEC共同主办。于是诞生了国际上最有影响力的美国APEC世界智能鼠竞赛。号称智能鼠世界三大赛事之一，截止到2019年已经举办了34届。

APEC组织的官网网址：http://www.apec-conf.org/。

美国智能鼠爱好者的网址：http://micromouseusa.com/，如图A-4所示。

（a）　　　　　　　　　　　　　　　（b）

图A-4　美国APCE世界智能鼠竞赛官网截屏

竞赛时间：每年2月到4月之间。

竞赛地点：每年不同（举办过的地点包括北卡罗来纳州、德克萨斯州、佛罗里达州、加利福尼亚州等），每年都会有来自美国、英国、日本、韩国、新加坡、印度、中国等国家的选手踊跃参赛，如图A-5所示。

图A-5　中国选手参加第30届美国APEC世界智能鼠竞赛

3. 英国智能鼠国际竞赛

从1980年至今，英国智能鼠国际竞赛已经成长为国际知名的智能鼠竞赛之一。

竞赛时间：每年6月。

竞赛地点：英国伯明翰城市大学。

该项竞赛由英国智能鼠和机器人协会主办，英国的智能鼠竞赛特点在于重在参与，从中学生、大学生、到社会人员，任何人都可以参赛。所有的参赛队员分为不同的组别，迷宫难度也适当调整。该项竞赛分为line follower、wall follower、maze solver等项目，吸引了来自全球10余个国家，50余支队伍参赛。

英国智能鼠国际竞赛评分规则介绍：在16×16的迷宫中，参赛智能鼠需要完成起点到终点的搜索和全迷宫的遍历，求解最佳路线并完成由起点到终点的

冲刺。计分时间=搜索时间（第一次搜索到终点的时间）/30+冲刺时间（完成起点到终点最短路径的高速冲刺）+惩罚时间（撞挡板罚时：3 s/次）。

官方网址：https://ukmars.org/index.php/Main_Page，如图A-6所示。

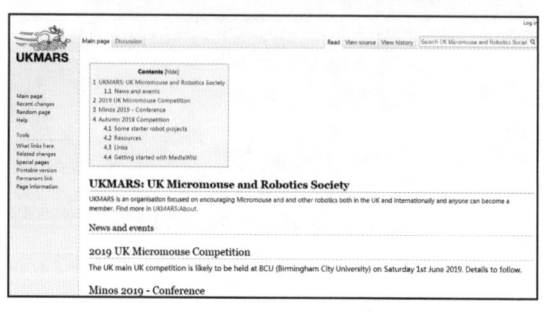

图A-6 英国智能鼠国际竞赛官网截屏

4. 全日本智能鼠国际公开赛

全日本智能鼠国际公开赛从1980年到2019年已经举办40届。

竞赛时间：每年的11月底或12月初。

竞赛地点：日本东京。

官方网址：http://www.ntf.or.jp/mouse/Micromouse2018/index.html，如图A-7所示。

视 频

全日本
Micromouse
国际公开赛

图A-7 全日本智能鼠国际公开赛官网截屏

每年竞赛都有来自美国、英国、日本、新加坡、中国、蒙古、智利、葡萄牙等二十多个国家的智能鼠参赛队角逐（见图A-8）。

赛项由古典智能鼠赛项、半尺寸智能鼠赛项、自走车赛项组成。参赛队由中学生、大学生和职业精英组成，据统计有300多支参赛队。全日本智能鼠国际公开赛可以说是代表当今，国际智能鼠技术领域级别最高、技术最强的赛事，所以备受瞩目。

图A-8　第39届全日本智能鼠国际公开赛颁奖照片

5. 智利智能鼠走迷宫国际竞赛

智利外交部希望通过国际智能鼠走迷宫竞赛，推动智利青少年科技创新以及国际间技术创新和交流合作，从而带动智利经济发展。2018年12月3日在全日本智能鼠国际公开赛期间，智利驻日本大使馆主持召开"智利智能鼠走迷宫国际竞赛研讨会"特别邀请国际智能鼠专家（美国的David Otten、英国的Peter Harrison、日本的中川友纪子、中国的宋立红、智利的Benjamin等）共同商议智利智能鼠走迷宫国际竞赛统一标准和规范，如图A-9、图A-10所示。

图A-9　智利外交部会议——共同探讨智能鼠发展

图A-10　2018年智利智能鼠走迷宫国际竞赛规则研讨会

6. 葡萄牙智能鼠走迷宫国际竞赛

2019年4月27日在Gondomar（贡多马尔）举办，由葡萄牙杜罗大学技术执行委员会主办。

葡萄牙竞赛开始于2011年，旨在通过培养创造力和能力来提供完整的技术学习环境，迄今已经成功举办9届。

竞赛时间：每年4月或5月。

竞赛地点：葡萄牙。

官方网址：http://www.micromouse.utad.pt/，如图A–11所示。

2019年4月27日，当地时间18时整，在葡萄牙波尔图体育馆中，来自英国、中国、葡萄牙、西班牙、巴西、新加坡等国家的参赛队，正在上演一场紧张激烈的国际智能鼠走迷宫竞赛。伴随着中国智能鼠稳健搜索和极速冲刺，掌声欢呼声在葡萄牙波尔图体育馆雷鸣般响起……启诚智能鼠实现突破性成果，取得了世界亚军的殊荣（见图A–12）。

视 频

葡萄牙
Micromouse
走迷宫国际
竞赛

图A–11　葡萄牙智能鼠走迷宫
国际竞赛官网截屏

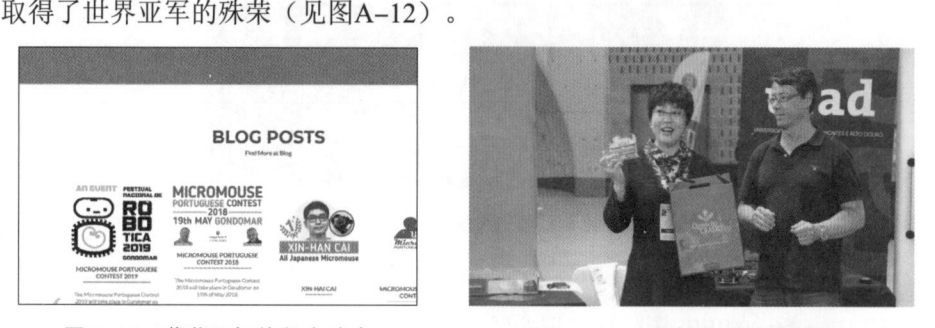

图A–12　启诚智能鼠荣获葡萄牙
智能鼠走迷宫竞赛世界亚军

赛后葡萄牙智能鼠走迷宫竞赛组委会主席安东尼奥表示，近年来中国的综合国力和技术实力不断增强，特别是教育领域对于科技创新和工程素养越来越重视。启诚智能鼠首次参加葡萄牙智能鼠走迷宫竞赛，就获得了优异成绩非常可喜可贺（见图A–13）。

图A–13　中国及葡萄牙智能鼠走迷宫竞赛专家现场技术交流

7. 印度智能鼠国际竞赛

2020年1月4日，印度孟买举办"2020年第23届亚洲科技节首届智能鼠国际大赛"，来自印度、中国、澳大利亚、尼泊尔、斯里兰卡、孟加拉等国家的代表队参加本届大赛（见图A–14）。中国天津智能鼠代表队力克群雄，以绝对优势包揽"金、银、铜"全部奖牌，将17.5万卢比奖金完美收入囊中。

竞赛时间：每年1月份。

竞赛地点：印度孟买。

官方网址：http://techfest.org/competitions/Micromouse。

● 视 频

印度
Micromouse
国际竞赛

图A–14　印度智能鼠国际竞赛合影

特别值得一提的是，印度金奈理工学院鲁班工坊代表队（见图A–15）采用2017年中方赠送的IEEE国际标准智能鼠走迷宫创新型教学设备TQD-Micromouse-JD智能鼠参加本届大赛，荣获印度国内大赛冠军、世界精英组第四名的好成绩，并赢得5 000卢比奖金，成为印度国际智能鼠走迷宫竞赛的明星赛队。印度鲁班工坊指导教师卡西克表示，金奈理工学院鲁班工坊代表队能取得这样优异的成绩，是三年来鲁班工坊的师生和支持企业（启诚科技）共同努力的成果。

图A–15　印度鲁班工坊参赛师生合影

8. 埃及智能鼠国际竞赛

埃及国际电气电子工程师学会（IEEE）现在已经发展成为具有较大影响力的国际学术和技术组织之一。30多年来，一直在推动和指导电气电子技术的发展与创新。这项技术包括电子元件、电路理论和设计技术的应用，以及针对有效转换、控制和电力状况分析工具的开发。IEEE成员包括杰出的研究人员、从业人员和杰出的获奖者。

图A-16所示为埃及IEEE在官网首页为智能鼠竞赛做的宣传。IEEE电力电子和可再生能源大会为颇具亮点的国际智能鼠大赛优胜者准备了丰厚的奖金。特等奖相当于1 000美元；杰出表现奖相当于700美元；最佳创新设计奖相当于500美元。参赛队来自埃及国内或国际工程学或相关专业的学生，也可以是高中生。每个参赛队中最多允许有两名学生。

官方网址：http://www.ieee-cpere.org/International_Competition.html。

图A-16　埃及智能鼠国际竞赛

埃及智能鼠国际竞赛纪实如图A-17所示。

图A-17　埃及智能鼠国际竞赛纪实照片

附录B

入门级智能鼠竞赛案例分析

IEEE国际标准智能鼠走迷宫竞赛具有一定难度，是一项富有挑战性和趣味性的学生比赛，在国内外享有一定的知名度和影响力。智能鼠走迷宫竞赛项目，从技术上涵盖了物联网应用技术、电子信息工程技术、嵌入式技术、通信技术、软件技术、计算机网络技术、信息安全技术、移动通信技术、计算机应用技术、应用电子技术、计算机控制技术、机电一体化技术、自动化技术等多个专业技术，涉及传感器检测、人工智能、自动控制和机电运动部件应用等技能和综合职业素养。全面展现高等教育和职业教育的发展水平，可提高电子信息类高素质、高技能应用型人才的培养质量。

为进一步深化普通中学教育改革，加强职教普教融合发展，探索多样化教育模式，天津市教育委员会，从2016年至2019年，连续四届举办天津市中等职业学校学生技能竞赛普职融合智能鼠走迷宫竞赛，将机械、电子、光学、自动控制、人工智能等多学科融合技术面向中学生进行普及推广，为普教和职教架起一座相互连通的桥梁。下面就以比较有代表性的普职融合智能鼠走迷宫竞赛的迷宫为范例做解析说明。

1. 2017年中等职业学校"普职融合"智能鼠走迷宫竞赛的迷宫解析

本次竞赛组队形式非常有创新性，每支参赛队都由2名中职生和1名中学生组成，竞赛内容包含理论知识考核和动手实际操作，中学生和中职生团结协作共同完成竞赛任务，如图B-1、图B-2所示。

图B-1　裁判对参赛队的实际操作任务进行考核

图B-2　裁判对参赛队的理论知识进行考核

本次竞赛所采用的8×8迷宫图非常经典，既有长直道可展现智能鼠的高速运动性能，也有多次连续转弯体现智能鼠的精确控制，如图B-3、图B-4所示。

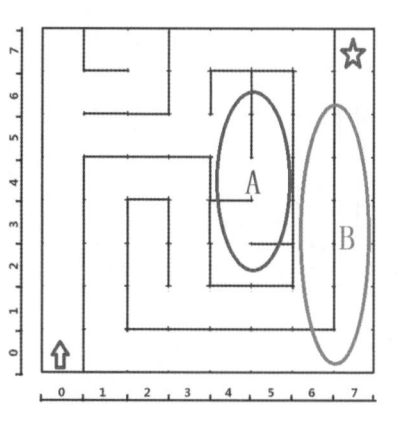

图B-3　最佳路径　　　　　　　图B-4　迷宫图解析

关键技术点A：考察学生对车体传感器检测和电动机运行参数的调试能力。智能鼠在迷宫中先后12次连续转弯运行，由于没有可用于矫正车姿的直道，所以对智能鼠的运行参数准确性有非常高的要求。

关键技术点B：因为是首届中学生参加的智能鼠竞赛，所以在迷宫图上设计了四个长直道，既有效降低了迷宫的整体难度，又能体现智能鼠的速度优势。

2. 2018年中等职业学校"京津冀"智能鼠走迷宫竞赛的迷宫解析

为了进一步落实"京津冀"教育协同发展，来自北京市、天津市、河北省参赛学校参加了本次竞赛。学生们综合运用传感、机电、自动控制等多方面的知识与技能，完成组装、调试、创新与竞速等任务（见图B-5）。

本次竞赛所采用的8×8竞赛迷宫，面向科普教育的基础学习者，对于学习入门级智能算法有很大的帮助，如图B-6、图B-7所示。

图B-5　2018年中等职业学校"京津冀"智能鼠走迷宫竞赛

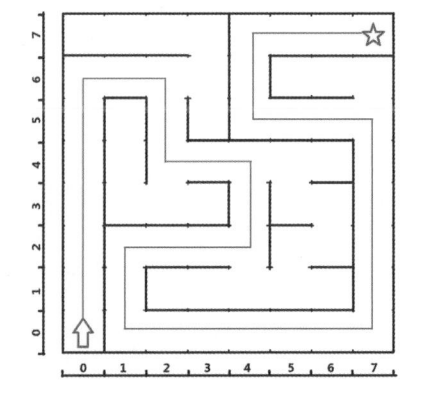

图B-6　最佳路径

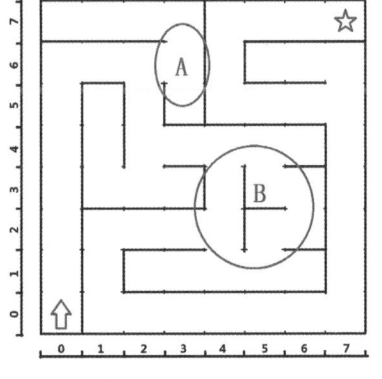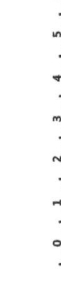

图B-7　迷宫图解析

关键技术点A：本次竞赛迷宫图针对智能鼠算法，设计了左手法则和右手法则的考核。技术点A处迷宫两侧均有路口，充分考察参赛者掌握基础智能算法的能力，因为每一次的转弯选择都很重要，如果选择错误，将影响到智能鼠后续能否顺利找到终点。

关键技术点B：该区域迷宫设计开放程度较高，在智能鼠搜索过程中六次无矫正连续转弯的考核，对培养参赛者智能基础算法的学习，起到促进和提升的作用。

3. 2019年中等职业学校职业技能大赛智能鼠走迷宫竞赛的迷宫解析

由来自天津17所职业学校和普通中学的学生参加此次竞赛（见图B-8）。场外观摩席观看竞赛的云南滇西地区红河哈尼族彝族自治州、怒江傈僳族自治州以及新疆和田的老师和学生们格外引人注目。带队新疆和田带队老师表示，在科技快速发展的大环境下，智能控制作为一种影响社会各个方面发展的颠覆性技术，对学生的生活和学习产生了重大影响，这些学生首次学习观摩智能鼠走迷宫竞赛，对于他们提升科技创新意识和动手设计能力非常有帮助，期待下

一届比赛，新疆和滇西的同学们也可以同场竞技，以赛促教、以赛促学，努力走出一条脱贫攻坚、职教帮扶的天津之路。

图B-8　参赛队学生们认真完成现场编程任务

本次竞赛所采用的8×8迷宫，考虑到观摩参赛队的学生的观看效果，在迷宫设计上突出了观赏性和学习性。除了有基础的智能算法即左手法则和右手法则的考核之外，还设计了难度比较高的"环形"字形迷宫路径，如图B-9、图B-10所示。

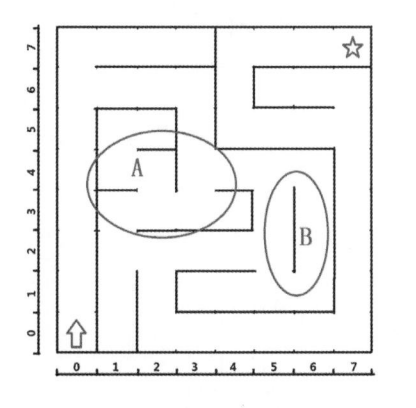

图B-9　最佳路径　　　　　　　图B-10　迷宫图解析

关键技术点A：该区域是智能鼠到达终点的路径之一，三次连续转弯对智能鼠的传感器检测精度和电动机运行稳定性提出了较高的要求。因为这段路较短，一般采用高级算法的智能鼠，通常会选择这段路程。

关键技术点B：该区域是另一条到达终点的路径，在迷宫设计上本着难度逐年递增的原则，特别设计了"环形"陷阱路径，对于初学者来讲，如果智能算法水平不够高，将很难顺利通过该路段。

4. 入门级智能鼠迷宫实训范例（见图B-11~图B-13）

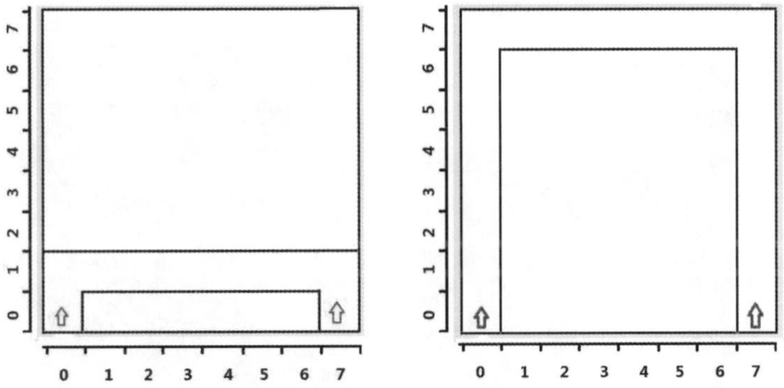

图B-11　初级练习迷宫图

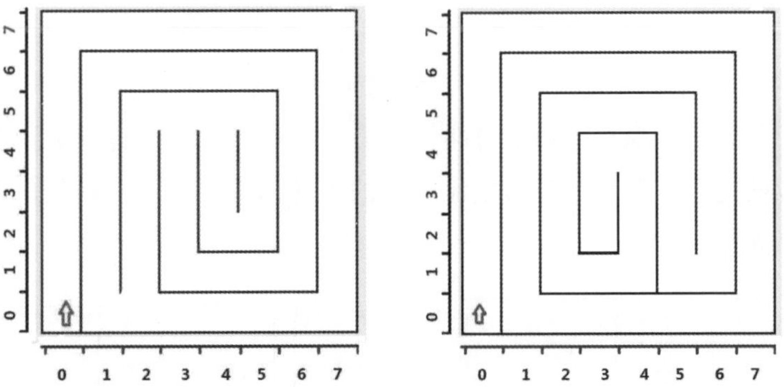

图B-12　能力提升迷宫图

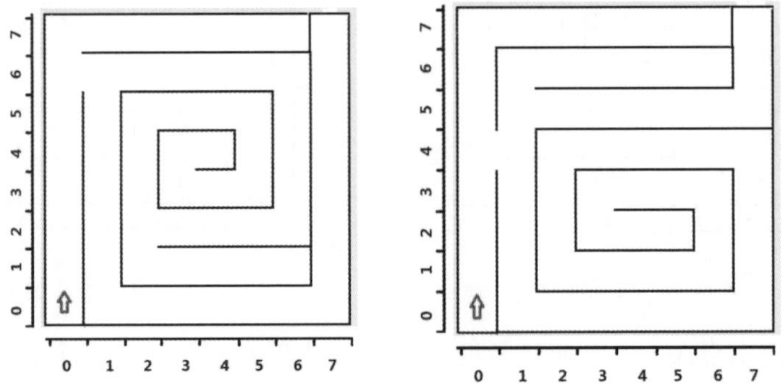

图B-13　高级竞赛迷宫图

附录C

TQD-Micromouse-JQ器件清单

TQD-Micromouse-JQ器件清单见表C-1。

表C-1 器件清单

序号	名称	数量	备注
1	TQD-Micromouse-JQ	1	
2	专用充电器	1	
3	USB线	1	
4	专用电池	1	
5	配套光盘	1	

附录D

TQD-IOT工程创新课程平台器件清单

TQD-IOT工程创新课程平台器件清单见表D-1。

表D-1 器件清单

序号	名称	数量	备注	序号	名称	数量	备注
1	小创客大智慧实验平台	1		13	无源蜂鸣器	1	
2	控制模块	1		14	红外对射	1	
3	蓝牙模块	1		15	人体红外	1	
4	液晶模块	1		16	振动模块	1	
5	单色LED I	1		17	风扇模块	1	
6	单色LED II	1		18	温度模块	1	
7	按键模块	1		19	流水灯模块	1	
8	三色LED	1		20	点阵模块	1	
9	光敏模块	2		21	MP3播放模块	1	
10	电位器模块	1		22	声控模块	1	
11	超声波模块	1		23	水滴模块	1	
12	有源蜂鸣器	1		24	舵机模块	1	

附录E

教学内容和学时安排

本课程参考教学学时为60学时，具体分配表见表E-1。

表E-1 学时分配

序号		内容	学时
第一篇	基础知识篇	项目一 智能鼠的发展历程 项目二 智能鼠的硬件结构 项目三 智能鼠的开发环境 项目四 智能鼠的基本功能控制	24
第二篇	综合实践篇	项目一 智能鼠的交互控制 项目二 智能鼠的姿态控制	14
第三篇	拓展竞技篇	项目一 智能鼠常用算法解析 项目二 智能鼠的高级控制	12
第四篇	拓展应用篇	项目一 IOT应用平台结构组成 项目二 智能鼠技术IOT扩展应用	10
总计			60

附录F

电路图形符号对照表

电路图形符号对照表如表F-1所示。

表F-1 电路图形符号对照表

序号	名称	国家标准的画法	软件中的画法
1	发光二极管		
2	电阻元件		
3	滑线式变阻器		

附录G

专业词汇中英文对照表

专业词汇中英文对照表见表G-1~表G-3。

表G-1 与智能鼠相关的专业词汇

中文	英文	中文	英文
核心控制模块	main control module	PWM信号发生器模块	PWM signal generator driver module
主控芯片	main control chip	左方、左斜、前方、右斜、右方	the left, the right, the front, the front-right and the right
输入模块	input module	g段	the g segment
输出模块	output module	空心杯直流电动机	coreless DC motor
核心板电路	main control circuit	步进电动机	stepping motor
电源电路	power circuit	电动机驱动电路	motor drive circuit
控制电路	control circuit	真值表	truth table
外围电路	peripheral circuit	H桥电路	H-bridge circuit
键盘显示电路	keyboard-display circuit	转动（步进电动机）	rotate
JTAG接口电路	JTAG interface circuit	电子元器件	electronic component
按键电路	key-pressing circuit	晶振	crystal oscillator
数据传输	data transmission	电容	capacitance
人机交互系统	human-computer interaction system	限流可调电阻	adjustable current-limiting resistance
吸地风扇技术	suction fan technology	数码管	digitron
占空比	duty cycle	外围器件	peripheral devices
角速度	angular velocity	脉冲振荡电路	pulse oscillation circuit
红外传感器	IR sensor	脉冲信号	pulse signal
红外检测电路	infrared detection circuit	方波	square wave
红外线	infrared light	感知系统	perceptual system
红外校准	infrared calibration	载波频率	carrier frequency
红外强度	infrared intensity	原理图	schematic diagram
红外发射头	infrared transmitter	软件界面	software interface
红外接收头	infrared receiver	驱动库	driver library

表G-2　与竞赛相关的专业词汇

中文	英文
单元格	cell
挡板	wall
立柱	post
竞赛场地	competition maze
起点	the start
目的地/终点	the destination
迷宫坐标	the coordinate in the maze
路口	crossing
电子自动计分系统	electronic automatic scoring system
参赛队员	competitor
智能鼠竞速比赛	micromouse competition
最优路径	the optimal path
轨迹	trajectory
通道	passage way

表G-3　与智能算法相关的专业词汇

中文	英文	中文	英文
底层驱动	the bottom driver program	差速控制	differential-speed control
顶层算法	the top algorithm program	直线运动	straight movement
算法	algorithm	转弯	turning
策略	strategy	校正车姿	correct the attitude
法则（左、右手法则）	rule	运行校正	attitude correction
右手、左手、中心法则	the right-hand rule, the left-hand rule, the central rule	核心函数	core function
90°、180°转弯	90-degree turning/180-degree turning	（驱动步进电动机的）时序状态	time sequence status
编程并实现	programming and realizing	前进一格	moving forward one cell
等高图	step map	按键等待	waiting for button press
循环检测	cycle detection	判断车姿	determining the attitude
"8字型"路径运行控制	movement control in picture-8-shaped path	暂停一步	waiting one step
实现避障	obstacle avoidance	精确转弯控制	accurate turning control
运动姿态的控制	motion attitude control	闭环控制	closed-loop control
两轮差速	two-wheel difference speed	绝对方向	absolute direction
路径规划和决策算法	path planning and decision algorithm	相对方向	relative direction
结构体	struct		

附录H

"智能鼠原理与制作" 国际实训课程标准

（适用于中等职业学校实训课程）

一、课程信息

授课对象：三年制中等职业教育层次学生。

适用专业：机电技术应用（051300）、电子技术应用（091300）、计算机应用（090100）、计算机网络技术（090500）、电气运行与控制（053000）。

课程类型：理论和实践。

建议开设学期：第三学期。

建议开设前导课程：电工电子技术及应用、机械制图/CAD、微机原理与接口技术、电子测量技术与仪器、C语言程序设计、单片机技术应用。

建议开设后续课程：传感器应用技术、电机技术、单片机原理及应用、自动控制原理、电子整机组装与调试。

二、课程性质

1. 符合教育部《中等职业学校专业教学国家标准》

"智能鼠原理与制作"从技术上涵盖了计算机应用、电子与信息技术、电子技术应用、机电技术应用、电气运行与控制等多种专业，涉及传感器检测、人工智能、自动控制和机电运动部件应用等技能和综合职业素养。课程设计与实施采用理实一体的方式，以动脑动手相结合，整体贯穿技术技能训练，为职业道德素养内化形成服务。

2. 对接国际化教学理念，服务技能型人才培养

在符合我国标准的基础上，本课程内容与IEEE国际竞赛内容对接，在教材课程设置中体现"应用性、项目化、任务式"的设计理念，采取国际流行的任务导向、基于工作过程系统化、工程实践创新等思想，让学生在学习专业知识的同时，通过小组教学、沟通协作、分阶段目标等形式提升学生的全面综合素养。

3. 服务"一带一路"倡议，推广中国教育标准

本课程内容与多个国家的"鲁班工坊"建设项目高度融合。作为中国教育的新名片，本课程服务"一带一路"倡议，推广中国教育标准，为"一带一路"沿线国家提供丰富实践教学资源，服务各地技术技能人才培养。

三、课程目标

本课程主要培养学生具备智能鼠系统结构分析与组装、集成开发环境搭建、传感与检测信号调试、智能鼠直行与转弯控制及姿态矫正、智能鼠智能搜索及路径规划的能力，探究能力和运用理论知识解决实际问题的能力。

1. 知识教学目标

（1）掌握智能鼠的基本硬件结构，了解装配、调试的基本原理与方法。

（2）了解智能鼠的核心控制板的控制原理。

（3）了解智能鼠的人机交互控制系统。

（4）掌握智能鼠常用算法的设计，了解实现智能搜索与路径规划的相关知识。

（5）了解智能鼠多传感器协同工作的原理。

2. 能力培养目标

（1）能够正确使用常用仪器仪表。

（2）具有查阅手册、阅读产品说明书等资料的能力。

（3）能够对智能鼠设备调试过程中的常见故障进行分析以及故障排除。

（4）具备探究能力和运用理论知识解决实际问题的能力。

（5）掌握智能鼠传感器和电动机的基础调试方法。

（6）掌握智能鼠的开发环境以及基本的使用方法。

3. 素质培养目标

（1）具有沟通及团队协作的能力。

（2）具有勇于创新、敬业、乐业的工作作风和严谨、求精、求实的工匠精神。

（3）具有安全意识、质量意识与责任意识。

四、内容和要求

建议总学时：60学时。

教学内容及要求见表H-1。

表H-1　教学内容及要求

序号	教学内容	教学要求	建议学时	重点或难点
1	第一篇　项目一　智能鼠的发展历程	了解智能鼠的起源，熟悉智能鼠的竞赛与调试环境	4	智能鼠的调试环境
2	第一篇　项目二　智能鼠的硬件结构	熟练掌握智能鼠硬件结构及各部分之间的相互关系；了解智能鼠关键器件的选型原则及方法；深刻理解智能鼠核心电路工作原理	8	智能鼠核心电路工作原理
3	第一篇　项目三　智能鼠的开发环境	熟悉Arduino开发环境，熟练掌握 智能鼠的程序下载方法	4	熟悉Arduino开发环境
4	第一篇　项目四　智能鼠的基础功能调试	让智能鼠的眼睛看世界，掌握智能鼠传感器的基础调节。让智能鼠的腿动起来，掌握智能鼠电动机的基础调节	8	传感器检测和电动机驱动
5	第二篇　项目一　智能鼠的交互控制	认识无线蓝牙的运行控制，掌握智能鼠眼睛和腿的协同工作	8	智能鼠眼睛和腿的协同工作
6	第二篇　项目二　智能鼠的姿态控制	让智能鼠跑起来，智能鼠学会转弯，掌握智能鼠电动机控制原理和驱动方法，完成智能鼠直行和转弯控制。掌握如何通过传感器修正智能鼠的动作姿态	8	智能鼠直行与转弯控制
7	第三篇　项目一　智能鼠常用算法解析	掌握左手法则、右手法则	4	迷宫搜索优化法则的原理
8	第三篇　项目二　智能鼠高级功能	掌握多传感器协同工作原理与实现，多线程工作原理与实现，智能鼠躲避障碍灵活运行	8	智能鼠灵活躲避障碍
9	第四篇　项目一　TQD-IOT工程创新课程平台结构组成	了解TQD-IOT平台和智能鼠的关系。了解TQD-IOT平台的硬件组成	2	TQD-IOT平台和智能鼠使用方法的区别
10	第四篇　项目二　智能鼠技术IOT扩展应用	掌握智能鼠控制器对灯光系统、安防系统和显示系统的控制方法	6	根据不同的应用完成对应的IOT实验

五、模式与方法

（1）教学模式要求：采用项目式教学模式，在工程背景下，实施工程实践导向、真实任务驱动式教学，其教学时采用实践载体智能鼠和情境IEEE迷宫真实现场、虚实结合、软硬结合的方式。实践过程要求"知技协进，德技并修"。要引导学生运用已学知识提高独立分析问题和解决问题的能力。

（2）授课教师的具体要求：本科及以上学历，中级以上职称，具有职业院校信息技术类、加工制造类专业的教学经历。

（3）实践教学环境条件：实训场地需地面平整、室内明亮、通风良好，场地面积应不小于100 m²，配套有实训工位、实训计算机、智能鼠平台、IEEE迷宫场地、拓展用电子信息类设备平台等，数量满足一个标准班（40人）的实训教学

需求。

（4）教学任务编制要求：项目任务书应明确教师讲授（或演示）的内容；明确学习者预习的要求；提出该项目整体安排以及各项目训练的时间、内容等。以小组形式进行学习，对分组安排及小组讨论（或操作）的要求，也应做出明确规定。

六、考核与评定

（1）考核方式：考查。

（2）成绩评定：理论和实际操作考核。

总成绩以百分制计算，分为平时成绩考核和期末综合考核两部分。平时成绩考核一般为出勤、项目作业、测验，占总成绩的 30%；期末综合考核一般为理论考试、实际操作考核，占总成绩的 70%。其中，理论考试占总成绩的 30%，实际操作考核占总成绩的40%。

七、教材

1. 推荐教材

王超，高艺，宋立红．智能鼠原理与制作：基础篇[M]．北京：中国铁道出版社有限公司，2019.

2. 参考资料

[1] 樊胜民，樊攀，张淑慧．Arduino编程与硬件实现[M]．北京：化学工业出版社，2019.

[2] 陈吕洲．Arduino程序设计基础[M]．北京：北京航空航天大学出版社，2015.

八、编写与审核

编写人：李　军　王晓芹　宋立红　冯　强　张家荃　陈立考　邱建国
审核人：龚　威　高　艺　贺东梅
2020年7月16日